PROCEEDINGS OF THE FOURTH INTERNATIONAL COLLOQUIUM ON
AGEING OF MATERIALS AND METHODS FOR THE ASSESSMENT OF
LIFETIMES OF ENGINEERING PLANT
CAPE TOWN / SOUTH AFRICA / 21-25 APRIL 1997

Ageing of Materials and Methods for the Assessment of Lifetimes of Engineering Plant

CAPE'97

Edited by
R.K.PENNY
Engineering Formation cc, Noordhoek, South Africa

A.A.BALKEMA / ROTTERDAM / BROOKFIELD / 1997

Photo cover: Darryl Roggen – Athlone Power Station, Cape Town, South Africa.
First commissioned in 1961.

The texts of the various papers in this volume were set individually by typists under the supervision of each of the authors concerned.

Authorization to photocopy items for internal or personal use, or the internal or personal use of specific clients, is granted by A.A. Balkema, Rotterdam, provided that the base fee of US$1.50 per copy, plus US$0.10 per page is paid directly to Copyright Clearance Center, 222 Rosewood Drive, Danvers, MA 01923, USA. For those organizations that have been granted a photocopy license by CCC, a separate system of payment has been arranged. The fee code for users of the Transactional Reporting Service is: 90 5410 874 6/97 US$1.50 + US$0.10.

Published by
A.A. Balkema, P.O. Box 1675, 3000 BR Rotterdam, Netherlands (Fax: +31.10.4135947)
A.A. Balkema Publishers, Old Post Road, Brookfield, VT 05036-9704, USA (Fax: 802.2763837)

ISBN 90 5410 874 6
© 1997 A.A. Balkema, Rotterdam
Printed in the Netherlands

Table of contents

Preface IX

Overview

Lifetime maximization for mechanical engineering problems 3
E. Schnack

Use of diagnostic system results for extension of lifetime of power plant 17
components
S. Vejvoda, D. Vincour & L. Růžek

Lessons to be learned from five unusual cases of high temperature material 23
degradation
M.J. Molyneaux & J.B. Speck

Life assessment techniques

Structural integrity assessment and lifetime prediction 39
B. Tomkins & N.M. Irvine

Decision-making for boilers with long-time operation 41
C. Delamarian & H.R. Kautz

Assessment of a transmission gas pipeline destruction 51
F. Valenta, J. Michalec, S. Konvičková, M. Růžička, J. Řezníček, M. Sochor &
M. Španiel

Technical basis and software development for flaw assessments 63
in NPP pipeline welds
A.I. Arzhaev, S.E. Bougaenko, I.N. Denisov, V.V. Aladinsky & V.O. Makhanev

Asset management

Risk based asset life management 71
J.R.Lilley & M.A.Davies

Probabilistic risk analysis of ageing components which fail on demand 85
A Bayesian model: Application to maintenance optimization of diesel engine
linings
C.A.Clarotti, A.Lannoy & H.Procaccia

Reliability analysis and safety evaluation on a nuclear power plant 95
Th.Meslin

Field application of 'Augur' ultrasonic system during RBMK NPP Unit ISI 97
and its impact on pressure boundary integrity
*A.I.Arzhaev, V.A.Kiselyov, V.G.Badalyan, A.Kh.Vopilkin, B.P.Strelkov,
V.N.Vanukov, V.A.Aladinsky & V.O.Makhanev*

Damage mechanics applications

The thermomechanical material state and integrity retaining of reactor vessel 107
under an anticipated accident
V.L.Danilov, M.V.Dobrov, S.V.Zarubin & Y.Fautrelle

Steel creep and creep rupture strength in environment containing hydrogen 113
V.L.Danilov & S.V.Zarubin

Gradual failure of trusses in creep conditions 117
M.Chrzanowski & P.Latus

Prediction of creep cracks in low alloy steel pipe welds by use 129
of the continuum damage mechanics approach
J.Storesund, P.Andersson, L.Å.Samuelson & P.Segle

Life extension of aged plant

Extending the reliable operation of ageing power stations through analysis 147
of failed components
H.C.Furtado, J.A.Collins & I.Le May

Lifetime assessment and repair of steam turbine casings and valve chests 157
K.H.Mayer, H.König, D.Weber & M.Weiss

Planning of power plant service and rehabilitation work – Reasons for service 167
life extension of older plants
H.R.Kautz

Life assessment methods for low alloy drum steels after long-term service 181
A.Hernas & L.Mirecki

Ageing effects

Under insulation corrosion – An inspection approach   193
A.W.Beattie

WWER-type nuclear reactor pressure vessel: Material radiation ageing issues   207
and effect of thermal annealing as a mitigation method
V.I.Levit

Assessment of operation-dependent structural life of ageing aircraft using   219
crack growth retardation model
J.-Y.Jeon & J.-Y.Lee

Physical ageing causes of alloy metals in the vibration conditions and method   229
for the phenomenon assessment
A.Jakowluk

Materials development

New ferritic steels increase the thermal efficiency of steam turbines   241
K.H.Mayer & H.König

9 to 12% creep resistant chromium steels for high pressure piping systems   251
in power plants
H.Weber & M.Zschau

Material development yesterday – today – tomorrow: Limitations of power   261
plant construction
H.R.Kautz

The susceptibility of low temperature sensitization of 304 S.S.   273
after mechanical stress improvement process (MSIP)
K.Y.Hsu

Fitness for service considerations: The Meyer hardness test applied to cold   285
rolled and annealed steel to analyse its physical state
Ph.Tipping & V.Levit

Non-metallic materials

Lifetime, toughness and reliability of engineering thermoplastics   297
A.Chudnovsky, D.Baron & Y.Shulkin

Fatigue investigation of polycarbonate used for aircraft canopies   309
H.Abramowitz, T.Hentea, Y.Kin & Y.Xu

Author index   318

Ageing of Materials and Methods for the Assessment of Lifetimes of Engineering Plant, Penny (ed.)
© *1997 Balkema, Rotterdam, ISBN 90 5410 874 6*

Preface

The subject of this Colloquium is of world-wide importance in the power generation field particularly. Not only is there a continuing and growing demand for electrical power but present and future supplies will inevitably have to come from utilities whose units have exceeded their design lives; for example, in the USA about 50% of all power produced is supplied by units which are 20 years old or older. This feature will continue – and also in petrochemical and other large installations – because capital expenditure can rarely be justified for new plant; as a rough guide, plant life extension costs are likely to be 10-20% of new installations per unit of output. Another reason for concentrating on life assessment and extension of plant operating at elevated temperatures is that the governing design codes of practice have proved to be of little help, other than in offering vague guidelines for component life. The code inadequacies stem from the fact that no account has been taken of time-dependency (ageing); for example, the ASME Boiler Code merely changes the value of the allowable stress in its routines for short-term, low-temperature operation and adds a requirement on the creep rate. Whilst such an over-simplified approach was thought to be conservative, its limitations have emerged in welded constructions in the petrochemical industry. Creep cracks have appeared in nozzles and other attachments due to combinations of high stress concentrations eroding margins of safety due to long time in service and, in some cases, reduced ductility under constrained conditions found in welds. Further problems may be expected in the future because of trends towards high operational temperatures where these effects will be amplified.

Several 'methodologies' have emerged over recent years which have been useful in bringing into focus the need for new approaches to plant life assessment. For example, it is essential that tools are available for non-destructive evaluations of components, which are proven in their abilities and also their limitations, so that rational analyses of defects can help in the reduction of periods between major shut-downs of plant. In addition, there have been raised optimisms regarding techniques for life assessment which are based on advanced computer techniques. In general though, the quest for the badly needed new approaches has given rise to large 'wish lists' for all sorts of materials data and, in this respect, it is perhaps worth harbouring the plea for pragmatism by L.N.Kachanov, whose innovative approaches to continuum damage are well known:

'...problems of damage mechanics are mainly engineering ones and ...it is essential to avoid superfluous formalism...'; and on materials data: *'...experiments in this field are difficult (especially under multi-axial stress and non-proportional loading). Therefore, experimental data as a rule are scarce. Determination of functions and constants which play a role in the complex variants of the theory from available experimental data is often practically impossible...'*

Those statements of Kachanov in 1986 are just as important today in view of the lack of resources (money and people) if methods are to be derived which are capable of practical application. This has been the main purpose of the CAPE International Colloquia series wherein interdisciplinary groups gather to keep abreast of trends in appropriate technologies; at the same time to try to distil results into useful procedures for use by practising engineers who have the huge responsibilities of operating and maintaining industrial plant at acceptable levels of safety and economy.

The present volume contains papers presented at CAPE'97 under the following main headings:

- Life assessment techniques;
- Asset management;
- Damage mechanics applications;
- Life extension of aged plant;
- Ageing effects;
- Materials development;
- Non-metallic materials degradation.

R.K. Penny
Editor

Overview

Lifetime maximization for mechanical engineering problems

E. Schnack
Institute of Solid Mechanics, Karlsruhe University, Germany

ABSTRACT: A very important question in the car industry, as well as in the aircraft and aerospace industry, is the question of the lifetime of machinery parts. It is remarkable that machinery parts are in general dynamically loaded. In the most important cases, we have high cycle fatigue behaviour as a result. Therefore, in this paper the question is analyzed of how the lifetime of constructions in mechanical engineering can be maximized by shape optimization.

1 INTRODUCTION

The lifetime of dynamically loaded machine constructions can be separated into two phases. The first one is the crack initiation phase and the second one is the crack propagation phase. In general, the crack propagation phase leads automatically to a loss of function of the machine construction. Therefore, it is more important to study the crack initiation phase. Analysts of the problem are therefore more interested in the maximization of the crack initiation phase, so that in the end a maximum lifetime of mechanical engineering constructions can be achieved. Moreover, the question if high cycle fatigue behaviour is important has to be discussed. Many constructions in the car industry and in the aircraft and airspace industry demand high cycle fatigue behaviour, so that this high cycle fatigue behaviour (HCF) is used to analyze the problem of maximization of the lifetime of machinery parts.

The idea of solving this problem by such a project is not new; it was first used in the 1950's, when some first hypotheses were defined in order to control this problem of maximization of lifetime for mechanical engineering constructions. In the literature of the years around 1950, one can find Neuber's hypothesis, Petersen's hypothesis and the statistical hypothesis based on a weakest link model, which was studies once more carefully together with numerical concepts (Fanni 1993). If these k_f (f stands for fatigue)-hypotheses are analyzed, one can see that the kernel parameter is the theoretical stress concentration factor (k_t) This led to the idea of controlling the stress concentration factor of critical machinery parts in order to achieve maximization of lifetime. Then some strategies were introduced to reduce the theoretical stress concentration factor, leading to a fully stressed design concept using shape optimization techniques. So for example there are some algorithms to control those problems with special optimization techniques. One of them is the fully stressed design concept (FSD) used with a special gradient free algorithm developed from the notch stress theory concept (Schnack 1979) (Schnack & Iancu 1989).

If this problem is studied carefully, introducing the k_f-hypotheses into the problem one will see that it is a very rough procedure, not considering for example the history of loading and the history of the plasticity development and the macro-cracking development in the machinery parts. So automatically the question arises of how to achieve by a more realistic concept a continuum in such a way that microcracking, plasticity and hardening are introduced so that the problem of maximization of lifetime is clarified for mechanical engineering problems.

Therefore - and this is very new compared with the literature at the moment - we are introducing the continuum damage mechanics (CDM) for the case of HCF. The field equations in CDM are very close to the non-linear equations for plasticity. Here micro-cracking and additionally plasticity will be considered, but plasticity does not occur on a meso- and macro-scale, but only on a micro-scale. That means there are small plasticity regions at each crack tip, but on the meso-scale there is no plasticity. Experiments have shown that due to the damage of the continuum only a linear behaviour can be observed. This will be considered in the HCF concept in the following. From that idea three cost functions are the result. The first one is a classical one, i.e. the minimization of maximum von Mises stress, leading to the FSD. The second one is the minimization of maximum damage equivalent stress, and the third one is the minimization of maximum damage itself.

Moreover, if this concept is used for practical problems in mechanical engineering, besides the definition of such damage continua by internal variables, different constraints, especially manufacturing constraints for the production technology have to be considered.

As for the numerical technique, it is used for realizing at first the non-linear equation system for the structure analysis by finite elements. To solve the free boundary value problem, free parameters are introduced to control the shape of parts of the structure. This can be done in a very simple way by finite element nodes, or more efficiently by B-spline approximation. Thus, B-spline nodes are used for the design variables. The optimization itself is done by solving a sequence of quadratic subproblems.

The theoretical and numerical results are compared with experiments using experimental equipment, especially a hydropulser machine. The specimens are formed by classical mechanical engineering methods, such as notch specimens of steel (30CrNiMo8) and they are loaded by sinusoidal tension/compression loads with the mean load 0. The experiments are carried out at a frequency of 20 Hz. If the lifetime of these parts is analyzed in the end, one will observe an increase by the remarkable factor of 3, compared with the classical cut-out notch specimens.

2 THE CONSTITUTIVE EQUATIONS FOR THE CONTINUUM DAMAGE MECHANICS (CDM)

As has been mentioned in the introduction, lifetime estimation by the k_f-hypotheses is only a very rough and heuristic model to be used for the maximization of lifetime of constructions in mechanical engineering. In particular, if complex loadings for

4

complex structures are considered, i.e. not only notches with single loadings, describing the problem by the k_f-hypotheses is not realistic. For this case, especially considering the loading history, working with CDM is a convenient tool (Kachanov 1986) (Lemaitre 1992).

Additionally, as was mentioned in the introduction, on the macro-scale there is no plasticity and on the micro-scale there is micro-cracking and plasticity clearly around the crack tips. Therefore a two-scale model has to be developed (Lemaitre & Chaboche 1990).

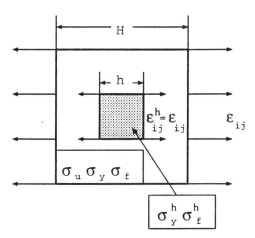

Figure 1. A two scale RVE

2.1 The two-scale model

Cutting out from the continuum representative volume element (RVE), grid parameters have to be introduced. As we are concerned with a two-scale model, two-scale parameters are needed (H for the macro-scale and h for the micro-scale) (see Fig. 1). For introducing continuum variables for the scale on the RVE, we define at first the ultimate stress (σ_u) and the yield stress (σ_y) for the fatigue limit:

$$\sigma_f \leq \sigma_y \tag{1}$$

Grid h is used for the micro-element assuming plasticity. The yield stress is defined by the fatigue limit of the given material:

$$\sigma_y^h = \sigma_f . \tag{2}$$

For the fatigue limit on the micro-scale, the relationship to the yield scales on the micro-scale is expressed in the following way:

$$\frac{\sigma_f^h}{\sigma_y^h} = \frac{\sigma_f}{\sigma_y}. \tag{3}$$

Considering Eq. 2, the following equation will be obtained for the fatigue limit on the micro-scale:

$$\sigma_f^h = \frac{\sigma_f^2}{\sigma_y}. \tag{4}$$

For the kinematical description of that model we are using the strain compatibility hypothesis Lin-Taylor (Taylor 1938), i.e. on the micro-scale there is the same stress tensor as on the macro-scale:

$$\varepsilon_{ij}^h = \varepsilon_{ij}. \tag{5}$$

2.2 The constitutive equations on the meso-scale

Because it is well-known that plasticity will occur from shear stresses and debonding of atomic structure from normal stresses, we are splitting up the stress and the strain tensor into the deviatoric and hydrostatic parts:

$$\sigma_{ij} = \sigma_{ij}^D + \sigma^{Hy}\delta_{ij} \qquad \text{and} \qquad \varepsilon_{ij}^e = \varepsilon_{ij}^{eD} + \varepsilon^{eHy}\delta_{ij} \qquad \text{with} \tag{6}$$

$$\sigma^{Hy} = \frac{1}{3}\sigma_{ll} \qquad \text{and} \qquad \varepsilon^{eHy} = \frac{1}{3}\varepsilon_{ll}^e$$

For that the strain energy is defined by the following:

$$w_e = \int \sigma_{ij}d\varepsilon_{ij}^e = \int \sigma_{ij}^D d\varepsilon_{ij}^{eD} + \int \delta_{ij}\delta_{ij}\sigma^{Hy}d\varepsilon^{eHy}. \tag{7}$$

As I have mentioned in the introduction, on the meso-scale only elastic behaviour can be observed, so the Hooke's law can be applied. Hooke's law is once more split up into the deviatoric and hydrostatic parts and, from that, the formulation for the effective stress and the strain equivalent formulation will follow:

$$\varepsilon_{ij}^{eD} = \frac{1+v}{E}\frac{\sigma_{ij}^D}{1-D} \qquad \text{and} \qquad \varepsilon^{eHy} = \frac{1-2v}{E}\frac{\sigma^{Hy}}{1-D} \tag{8}$$

In this approach the damage variable on the meso-scale will remain 0 ($D = 0$ in Eq. 8), until the micro-element fails, i.e. $D = 1$. From that follows that we have $D = 1$ on the macro-scale. A stress redistribution in this formulation only occurs on the micro-scale (Lemaitre 1992). In comparison to that (Grunwald 1996), one can find that there can be a coupling of damage on the micro- and the meso-scale. With Eqs. 7

and 8 we will obtain the following for the strain energy and consequently a special norm of the stress tensor in the formulation of the von Mises equivalent stress:

$$\sigma_{eq} = \left(\frac{3}{2}\sigma^D_{ij}\sigma^D_{ij}\right)^{\frac{1}{2}}. \tag{9}$$

The associate variable to the damage variable is the strain energy density release rate, similar to the one we have in classical fracture mechanics:

$$Y = \frac{w_e}{1-D} = \frac{\sigma^2_{eq}}{2E(1-D)^2}\left[\frac{2}{3}(1+v)+3(1-2v)\left(\frac{\sigma^{Hy}}{\sigma_{eq}}\right)^2\right]. \tag{10}$$

$\dfrac{\sigma^{Hy}}{\sigma_{eq}}$ is called the triaxiality function. The kernel of the energy release rate is the triaxiality function R_v :

$$R_v = \frac{2}{3}(1+v)+3(1-2v)\left(\frac{\sigma^{Hy}}{\sigma_{eq}}\right)^2. \tag{11}$$

The damage equivalent stress σ^* can be found if we define a relationship between the strain energy for the one-dimensional case - i.e. $R_v = 1$ - and for the three-dimensional situation:

$$w_e = \frac{\sigma^{*2}}{2E(1-D)} = \frac{\sigma^2_{eq}}{2E(1-D)}R_v. \tag{12}$$

From that follows for the damage equivalent stress σ^* :

$$\sigma^* = \sigma_{eq}R_v^{\frac{1}{2}}. \tag{13}$$

2.3 The Constitutive Equation on the Microscale

At first an initial for the damage evolution law is needed. In order to find this some experiments (Lemaitre 1992) have been carried out. This leads as a result to the following rate formulation on the micro-scale for the damage variable:

$$\dot{D}^h = \frac{Y^h}{S}\dot{p}^h\Theta(p-p_D) \qquad \text{with} \qquad \Theta(p-p_D) = \begin{cases} 1 & \text{if: } p \geq p_D \\ 0 & \text{else.} \end{cases} \tag{14}$$

In Eq. 14 S is the damage strength, which can be interpreted as a material constant. The dominant variable for controlling the damage increment is the strain energy release rate Y^h. p^h is a norm of the accumulated plastic strain:

$$\dot{p}^h = \left(\frac{2}{3} \dot{\varepsilon}_{ij}^{ph} \dot{\varepsilon}_{ij}^{ph} \right)^{\frac{1}{2}}.$$
(15)

Now the yield criteria can be formulated as well known from the von Mises theory, but now outlined for the microscale:

$$f = \frac{\sigma_{eq}^h}{1-D} - \sigma_y^h = \frac{\sigma_{eq}^h}{1-D} - \sigma_f = 0,$$
(16)

For the damage constitutive equation we obtain as a result:

$$\dot{D}^h = \frac{\sigma_f^2}{2ES} R_V^h \dot{p}^h \Theta(p - p_D).$$
(17)

As can be seen in Eq. 14, the damage increment is directly proportional to the increment in accumulated plastic strain. Plastic straining in high cycle fatigue behaviour is only observed on a microscale. I repeat this to underline that there is no plasticity on the meso-scale. Therefore, the damage evolution is also related to the micro-scale.

We now come to the experimental part. The comparison with a tension test determines:

$$p_D^h = \frac{\sigma_u - \sigma_f}{\sigma_f - \sigma_f^h} \varepsilon_{pD}$$
(18)

where ε_{pD} is a plastic strain related to the ultimate stress in a uniaxial tension test. Altogether the constitutive equations are now formulated for the microscale in the following set:

$$\varepsilon_{ij}^h = \varepsilon_{ij}^{eh} + \varepsilon_{ij}^{ph},$$

$$\varepsilon_{ij}^{eh} = \frac{1+\nu}{E} \frac{\sigma_{ij}^h}{1-D} - \frac{\nu}{E} \frac{\sigma_{ll}^h}{1-D} \delta_{ij},$$

$$\dot{\varepsilon}_{ij}^{ph} = \begin{cases} \frac{3}{2} \frac{\tilde{\sigma}_{ij}^{hD}}{\sigma_f} \dot{p}^h & \text{if:} \quad f = 0 \wedge \dot{f} = 0, \\ 0 & \text{else,} \end{cases}$$

$$\dot{D} = \begin{cases} \frac{\sigma_f^2}{2ES}R_V^h\,\dot{p}^h & \text{if:} \qquad p^h \geq p_D^h \wedge f = 0 \wedge \dot{f} = 0 \\ 0 & \text{else.} \end{cases} \tag{19}$$

3 COMPUTATIONAL ASPECTS FOR THE STRUCTURAL ANALYSIS

The finite element method is used with triangular elements formulated with linear shape functions. As I have mentioned in the introduction, the continuum damage mechanic algorithm works in principle like the elasto-plasticity carried out with finite elements. The actual time step will be indicated in the following by $k + 1$. Because the variables are formulated in the following concept on the micro-scale, we cancel the index h. Only the fatigue limit σ_f belongs to the meso-scale. All the time dependent variables for the following step $k + 1$ are formulated without notation to $k + 1$.

After these definitions an elastic increment for the predictor can be started:

$$\widetilde{\sigma}_{ij} = \frac{Ev}{1-2v}\varepsilon_{ll}\delta_{ij} + \frac{E}{1+v}(\varepsilon_{ij} - {}^k\varepsilon_{ij}^p), \tag{20}$$

If the elastic predictor satisfies the condition $f \leq)$, to elastic predictor was correct. If the yield condition is violated by $f > 0$, a plastic correction is necessary. For the integration scheme the following set is applied, where f defines the yield condition:

$$f = \widetilde{\sigma}_{eq} - \sigma_f = 0,$$

$$\widetilde{\sigma}_{ij} = \frac{Ev}{1-2}\varepsilon_{ll}\delta_{ij} + \frac{E}{1+v}(\varepsilon_{ij} - {}^k\varepsilon_{ij}^p - \Delta\varepsilon_{ij}^p),$$

$$\Delta\varepsilon_{ij}^p = N_{ij}\Delta p,$$

$$\Delta D = \frac{Y}{S}\Delta p, \qquad \text{with} \tag{21}$$

$$N_{ij} = \frac{3}{2}\frac{\sigma_{ij}^D}{\widetilde{\sigma}_{eq}}.$$

For the unknown variables $\widetilde{\sigma}_{ij}$ and p there are two non-linear equations:

$$f = \widetilde{\sigma}_{eq} - \sigma_f = 0 \tag{22}$$

and

$$h_{ij} = \widetilde{\sigma}_{ij} - \frac{Ev}{1-2v}\varepsilon_{ll}\delta_{ij} - \frac{E}{1+v}\left[\varepsilon_{ij} - {}^k\varepsilon_{ij}^p) - N_{ij}\Delta p\right] = 0. \tag{23}$$

9

For the solution, Newton's method is used. As a result we get an iteration scheme with two equations:

$$f + \frac{\partial f}{\partial \tilde{\sigma}_{ij}} C_{ij}^{\tilde{\sigma}} = 0 \tag{24}$$

and

$$h_{ij} + \frac{\partial h_{ij}}{\partial \tilde{\sigma}_{pq}} C_{pq}^{\tilde{\sigma}} + \frac{\partial h_{ij}}{\partial p} C^p = 0 \tag{25}$$

where f, C_{ij} and the coefficients defined by partial derivatives are taken at iteration s for the timestep $k + 1$. The corrections are achieved by the following set of equations:

$$C_{ij}^{\tilde{\sigma}} = {}^{s+1}\sigma_{ij} - {}^s\sigma_{ij} \tag{26}$$

and

$$C^p = {}^{s+1}p - {}^sp \tag{27}$$

$$C^p = \frac{2(1+v)(f - N_{ij}h_{ij})}{3E} \qquad \text{and} \tag{28}$$

$$C_{ij}^{\tilde{\sigma}} = -\frac{2}{3}(f - N_{qp}h_{qp})N_{ij} - \frac{h_{ij} + \dfrac{E\Delta p}{\sigma_{eq}(1+v)3} N_{qp}h_{qp}N_{ij}}{1 + \dfrac{3}{2}\dfrac{E}{\tilde{\sigma}_{eq}(1+v)}\Delta p} \tag{29}$$

After solving this scheme we have as a result $\tilde{\sigma}_{ij}$ and p, where ε_{ij}^p and D are calculated by the discretized constitutive equations with the post-processing. The stress components can be computed then:

$$\sigma_{ij} = (1-D)\tilde{\sigma}_{ij}. \tag{30}$$

To analyze the structure for each cycle step by step (up to 10^6 - 10^7 cycles) is impossible, because a very CPU-time would be necessary. Therefore a jump procedure is defined, i.e. partly in an interval a linear behaviour is assumed within the next ΔN cycles for the increment in plastic strain and damage accumulation. This will reduce the CPU-time drastically, so that the iterative scheme for the elasto-plastic continuum damage mechanics can be realized as a realistic computation.

4 THE OPTIMIZATION PROCEDURE

By generalizing the problem, we first consider a cost function depending on design parameters t:

$$\mathcal{F} = \min f(t) \qquad t \in \Re^n. \tag{31}$$

There is a set of equality and inequality constraints:

$$h_i(t) = 0 \qquad \text{for} \qquad i = 1,\dots,m_h \qquad \text{and} \tag{32}$$

$$g_i(t) \le 0 \qquad \text{for} \qquad i = 1,\dots,m_g \tag{33}$$

with constraints to the design vector:

$$\underline{t_i} \le t_i \le \overline{t_i} \qquad \text{for} \qquad i = 1,\dots,n \tag{34}$$

The constraint optimization problems can be transformed into an unconstraint formulation by the Lagrange function:

$$L(t,\lambda,u) = f(t) + \lambda_i g_i(t) + \mu_j h_j(t) \qquad \text{with} \qquad i = 1,\dots,m_g, j = 1,\dots,m_h \tag{35}$$

and: λ, μ: Lagrange multipliers.

The optimization problem is solved by the algorithm (SQP) (Schittkowsky 1984). A very important point is to outline the cost functions for the high cycle fatigue problem. As I have explained, one can see that in HCF there is no plasticity on a macro-scale. This leads to the first cost function of the problem using the von Mises equivalent stress:

$$\mathcal{F} = \min f_1(t) \ \min(\max_{i=1,\dots,m} \sigma_{eq}^i) \ \text{with } m : \text{number of nodes on the free boundary.} \tag{36}$$

The definition of Eq. (36) is realized by Finite Elements, thus a discretised form. Therefore $f_1(t)$ is non-differentiable, and we will have to work with a transformation with an additional design variable β, which is used as cost function as well:

$$\mathcal{F} = \min \beta \tag{37}$$

(Schnack and Spörl 1986)

With this transformation we obtain an additional set of constraints:

$$\sigma_{eq}^i - \beta \le 0 \qquad I = 1, \dots, m \tag{38}$$

More precisely, we are now introducing the second cost function, going back to the characterizing fatigue behaviour of this problem. The second cost function minimizes the maximum damage equivalent stress, which has been defined in the previous chapter:

$$\mathcal{F} = f_2(t) = \min(\max_{i=1,\dots,m} \sigma_i^*).$$ (39)

I repeat that on a macroscale, there is no difference between the von Mises stress and the damage equivalent stress, if we analyze two-dimensional structures. But for a microscale the difference is significant, due to plasticity and incompressibility of the material, so that in the end a significant difference in the cost function can be observed compared with Eq. 36.

The third cost function is defined by the maximum damage at a given number of load cycles. Here we are looking for the minimum of the maximum damage at a given number of load cycles and we get

$$\mathcal{F} = f_3(t) = \min(\max_{i=1,\dots,m} D_i).$$ (40)

The number of cycles is determined by the following equation:

$$N - N_{Dc} = 0$$ (41)

We have to note that N_{Dc} is the data of the statically optimized shape with failure for the given load. This leads us to the definition of the critical damage $(D = D_c)$.

Besides the cost functions constraints have to be considered. At first the free boundary must be in the variation domain Γ^*, leading to:

$$(x_i, y_i) \in \Gamma^* \qquad i = 1, \dots, m$$ (42)

This leads us to restrictions of the coordinates of the free boundary:

$$\underline{x_i} \le x_i \le \overline{x_i} \qquad \text{and}$$ (43)

$$\underline{y_i} \le y_i \le \overline{y_i}$$ (44)

A similar formulation can be realized, if the design variables are not the nodes of the coordinates of the Finite Elements, e.g. defined by de-Boor ordinates of a B-spline approximation.

The next step is to introduce constraints to avoid mesh degeneration of the Finite Element network. For that we have to consider:

$$\sqrt{(x_i - x_{i-1})^2 + (y_i - y_{i-1})^2} - \sqrt{(x_{i+1} - x_i)^2 + (y_{i+1} - y_i)} = 0$$ (45)

with $\qquad\qquad i = 2, \dots, m-1$

In mechanical engineering problems it does not make sense to have e.g. concave and convex curvatures along the free boundary. Therefore we are introducing (Fanni

1993) a concave constraint which is formulated for the discretized structure with the following determinant:

$$\begin{vmatrix} 1 & 1 & 1 \\ x_{i-1} & x_i & x_{i+1} \\ y_{i-1} & y_i & y_{i+1} \end{vmatrix} = 2A_i \geq 0 \tag{46}$$

Finally it is important to note that the mechanical engineering structure must be producible, and therefore a minimum curvature radius of the cutter radius of the CNC milling-cutting machine is introduced:

$$R_{min} - R_i \leq 0 \qquad i = 1, \ldots, m - 1 \tag{47}$$

Thus the structure analysis for the continuum damage mechanics has been carried out, as well as a detailed description of the optimization technique in order to solve the problem of maximizing the lifetime of machinery parts for the case of high cycle fatigue behaviour.

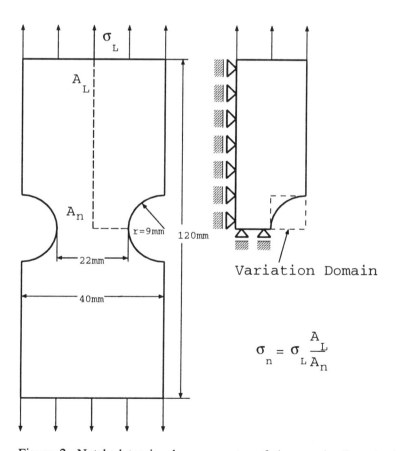

Figure 2. Notched tension bar, a quarter of the part is discretized because of the inherent symmetry.

5 NUMERICAL AND EXPERIMENTAL TESTS

For the maximization of the lifetime of mechanical engineering structures we have a lot of experience to control the equivalent von Mises stress value, i.e. to reduce the maximum of the von Mises stress value to a minimum. From that we are now starting to test our new theoretical model in shape optimization for high-cycle fatigue behaviour. As a test object a notched tension bar is used, as can be seen in Fig. 2, where the classical cut-out like a circle and the defined variation domain are depicted.

A typical Finite-Element mesh can be seen in Fig. 3. In order to define these structures by Finite Elements and to control the equivalent von Mises stress value, results can be found in some papers (Schnack 1979)(Fanni et al 1994).

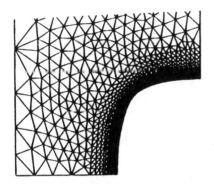

Figure 3. FE-mesh of an optimized part with adaptive substructure.

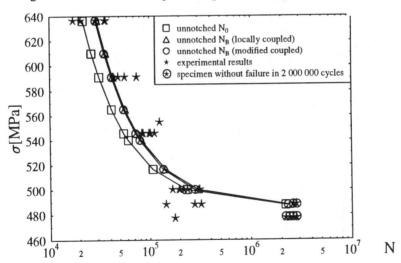

Figure 4. Experimental data and numerical S-N-diagrams for 30CrNiMo8 (Grunwald 1996). N_0 : start of damage evolution; N_B : failure of the specimen.

We are now starting with shape optimization for high-cycle fatigue behaviour. therefore we are defining the free boundary with 9-de-Boor-ordinates as design variables and for that we have 37 FE-nodes. Tests controlled by error estimation have shown that this starting profile is not good enough. A refinement of the structure is necessary so that we come up in the end to 65 nodes on the free boundary. The manufacturing constraints for producing this notched bar were cut for a milling-cutting machine with a minimum radius of R_{min} = 3mm. For the lifetime experiments all specimens were polished and annealed by 570°C after milling and air-cooled during the experiments. The load was a sinusoidal tension-compression with a frequency of 20 Hz, whereby the mean load was zero. In Fig. 4 the numerical Woehler lines (S-N-diagram) for an unnotched specimen are compared with experimental results. These is a good agreement between the experimental data by this diagram. In Fig. 5 the Woeler line is constructed for a circularly notched, the statically optimized, the σ*-optimized and the "damage optimized" results.

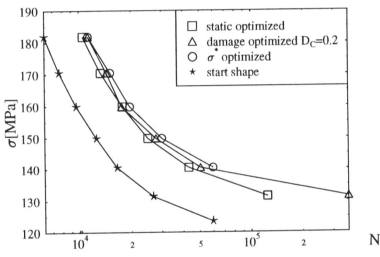

Figure 5. Numerical S-N-diagrams for the circularly notched parts and the parts with optimized shapes. Material: stainless steel at room temperature (Lemaitre 1992).

The difference between the lifetimes of the statically optimized and the "damage optimized" results is small. From that we can see the consistence between the three used cost functions defined before. It is very important to note that the lifetime of the structure optimized by the new method can be enlarged by three times in comparison with the classical circularly notched specimen.

6 CONCLUSION

In this paper we can see a new method based on the continuum damage mechanics for the maximization of the lifetime of dynamically loaded machine parts and the construction with a high cycle fatigue behaviour. Besides the structure analysis model with the two-grid formulation, the optimization procedure is outlined, realized together with structure analysis in Finite-Element notation. The experimental results show a significant fact in the maximization of lifetime. For the specimen used in the test, three times longer lifetimes were achieved for notched structures compared with the ones with classical cut-outs.

ACKNOWLEDGEMENTS

We would like to express our appreciation to the German Research Foundation (DFG) for financial support in the form of the projects Schn 245/15-1, Schn 245/15-2 and Schn 245/15-3.

REFERENCES

Fanni, M. 1993. *Formoptimierung dynamisch belasteter Bautiele,* PhD Thesis, University of Kralsruhe.

Fanni, M., Schnack, E. and Grunwald, J. 1994. Lifetime maximization through shape optimization of dynamically loaded machine parts. *Int. J. for Eng. Analysis and Design,* Vol. 1; pp 25-41.

Grunwald, J, and Schnack, E. 1995. *Formoptimierung dynamisch belasteter Bautiele.* DFG Final Report 1.9.1991 - 31.12.1995, DYN-OPT, Schn 245/15-1, 15-2, 15-3.

Grunwald, J. 1996. *A fatigue model for shape optimization based on continuum damage mechanics,* PhD Theses, University of Karlsruhe.

Grunwald, J. and Schnack, E. 1996. Shape Optimization of Damaged Continua Considering HCF. *Proc. of TMCE,* Budapest, pp 75-89.

Grunwald, J. and Schnack, E. 1997. A Fatigue Model for Shape Optimization. *Structural Optimization,* Springer Verlag, to appear.

Kachanov, L.M. 1996. *Introduction to continuum damage mechanics,* Martinus Nijhoff Dordrecht, The Netherlands.

Lemaitre, J., and Chaboche, J.-L. 1990. *Mechanics of solid materials,* Cambridge University Press.

Lemaitre, J. 1992. *A course on damage mechanics,* Springer Verlag, Berlin, Heidelberg and New York.

Lemaitre, J, and Doghri, J. 1994. Damage 90, a postprocessor for crack initiation, *Computer Methods in Applied Mechanics and Engineering.* Vol. 115, n 3-4 pp 197-232.

Schittkowski, K. 1984/85. NLPQL, A fortran subroutine solving constrained non-linear programming problems. *Annals of Operation Research,* 5, pp 485-500.

Schnack, E. 1979. An optimization procedure for stress concentrations by the finite element technique, *Int. J. Numer. Meth. Eng. 14,* 1, pp 115-124.

Schnack, E., and Spörl, U. 1986. A Mechanical Dynamic Programming Algorithm for Structure Optimization. *Int. J. Numer. Meth. Eng. 23,* 11, pp 1985-2004.

Schnack, E., and Iancu, G. 1989. Shape design of elastostatic structures based on local perturbation analysis, *Structural Optimization,* 1, pp. 117-125.

Taylor, G.I. 1938. Plastic strain in metals, *J. Inst. Met.,* pp 62-307.

Ageing of Materials and Methods for the Assessment of Lifetimes of Engineering Plant, Penny (ed.)
© *1997 Balkema, Rotterdam, ISBN 90 5410 874 6*

Use of diagnostic system results for extension of lifetime of power plant components

Stanislav Vejvoda, Dušan Vincour & Leoš Růžek
Vítkovice Institute of Applied Mechanics Brno, Ltd, Czech Republic

ABSTRACT: The power plant components are manufactured for design conditions. Calculation of temperature fields, strains, stresses and assessment of limit states (static, fatigue, creep, brittle crack, resistance to influence of corrosive environment) are done for design conditions. The significant deviation of actual service conditions from design ones are found out by diagnostic systems. Results of the diagnostic system evaluation can be utilized for the controlled ageing and lifetime extension of power plat components. The diagnostic system DIALIFE can be utilized for this purpose.

1 DESIGN SERVICE CONDITIONS

Power plant components are designed and manufactured for design service conditions. These conditions can be based on theoretical analyses of processes for which components are determined, or on analyses of experimental measurement of similar power plant components. The designer tries to describe the service of components with the best accuracy, but he cannot know all details of the actual service. The design conditions of the WWER power plant units as an example are shown in the Fig. 1, (Lička 1994). The temperatures and pressures of the media for start-up, nominal and shutdown of the unit are there expressed as time dependent. These relations are usually linear ones. The transient conditions are described similarly. These usually start and finish on the nominal conditions. The cases exist, only when nominal conditions and number of unit start-up and shutdown are known.

2 ACTUAL SERVICE CONDITIONS

The detailed information about actual service conditions of power plant components enables to obtain information and control systems (I and C), diagnostic systems (DS) and special temporary measurement (STM). The possibility of the

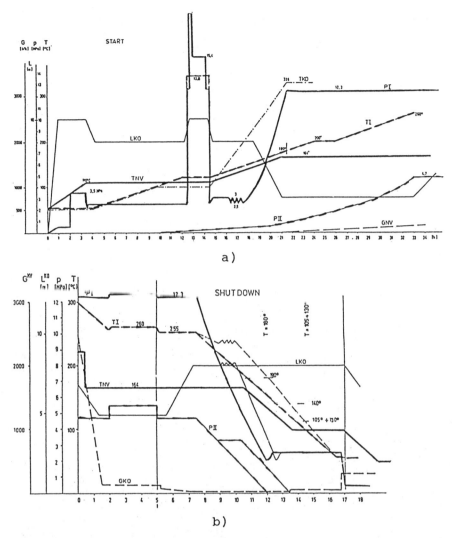

Figure 1. Design service conditions of power plant unit, a) start-up, b) shutdown, PI, TI - pressure and temperature of the primary circuit, PII - pressure of the secondary circuit, TNV - feed water temperature.

OFF-LINE diagnostic system DIALIFE are described by Vejvoda, Moulis and Vinçour (1996). As sensors thermocouples, strain ganges, accelerometers, sensors for displacement measuring etc. are used.

Results of temperature measurement on the outside surface of the steam generator feed nozzle, temperature of the feed water and the secondary medium during start-up of the unit are shown in the Fig. 2a, during shutdown on the Fig. 2b and during one transient condition on the Fig. 3. The results show that every start-up, shutdown and the same transient event are different at every repeating.

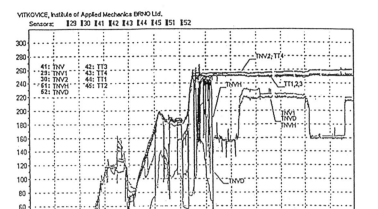

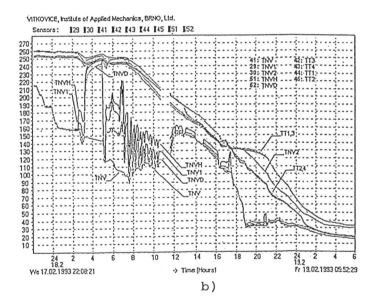

Figure 2. Temperatures on the outside surface of the steam generator, a) start-up, b) shutdown of unit. TNV - feed water, TNV1, TNVH, TNVD - feed water nozzle (H - upper, 1-middle and D - lower surface line), TNV2 - outside surface of steam generator close to the nozzle, TT1,2,3,4 - upper and lower surface line of the steam generator.

Actual changes of temperatures versus time are digressed from design ones, compare the Fig. 1 with the Fig. 2. It means that the actual damage of the component material during service is different from the damage calculated for design service conditions.

19

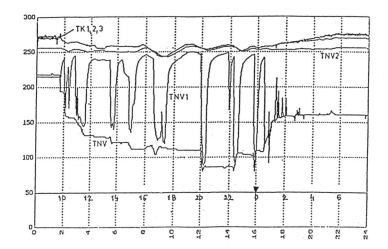

Figure 3. Actual service conditions of one transient event.
Temperatures on the outside surface of the steam generator,
TNV - feed water, TNV1 - feed water nozzle, middle surface
line, TNV2 - outside surface of steam generator close to
the nozzle, TT1,2,3 - upper and lower surface line of the
steam generator.

Data measured by the I and S, DS or STM can show, that
service limits were over- or under-cross. For example the
feed water temperature during normal service condition has
to be 164 °C. The Fig. 3 shows, that this temperature (TNV)
was less that required. The decrease of the water temperature is on behalf of the interruption of the water flow
when power is close to zero. The water is stopped in the
long part of the thermal uninsulated piping. Every interruption of the water flow causes a stress cycle and fatigue
of the feed water nozzle material. These changes of stresses were not taken into consideration in the design analysis.

All processes were analysed and very simple relation was
proposed for the approximate evaluation of the damage D of
feed water nozzle material :

$$D = \sum_{i=1}^{12} \frac{n_{Ti}}{[N_o]_i} = \frac{n_{205}}{2686} + \frac{n_{195}}{3114} + \frac{n_{185}}{3683} + \frac{n_{175}}{4484} + \frac{n_{165}}{5578} + \frac{n_{155}}{7120} +$$

$$+ \frac{n_{145}}{9349} + \frac{n_{135}}{12667} + \frac{n_{125}}{17825} + \frac{n_{115}}{26168} + \frac{n_{105}}{36265} + \frac{n_{95}}{51184} \ ,$$

(1)

where
n_T - number of the occurrence of the temperature difference $T = T_{II} - T_{NV}$ during a whole service time of the
unit, where T_{II} is temperature of the secondary medium and T_{NV} is temperature of the feed water,

20

[N_o] - allowable number of the occurrence of the T, which itself causes a total damage of the feet water nozzle material.

The design changes of the feeding regulation and the thermal insulation of the piping makes possible to decrease the damage of the feed water nozzle material. These measures are necessary where the extension of lifetime of power plant components is required.

Similar diversion of actual service conditions from design ones has been observed at the other parts of power plant component, for example water spray nozzle, horizontal parts of pipelines etc.

The OFF-LINE system DIALIFE is installed on boilers too, Fig. 4. Result of the creep analyses using of measured temperatures of the material and temperature and pressure of medium will used for the proposals extension of the lifetime of the boiler.

The high level of bending stresses in stud bolts was observed by strain gauges, when their tightening by the torque wrench was used. The average stress in the cross section of the tightened stud bolt has to be equal 220 MPa. The parasitic bending stress was measured in range from 150 MPa

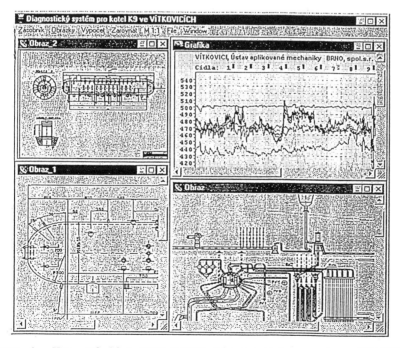

Figure 4. Use of the OFF-LINE diagnostic system DIALIFE for the boiler K9 in the VITKOVICE company. The scheme of the boiler details and measured temperatures on the outside surface of the boiler K9 demonstrated on the PC display.

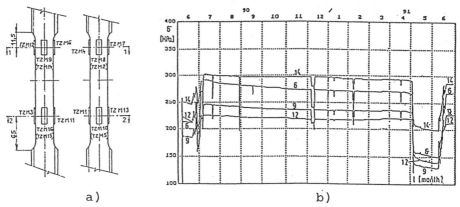

Figure 5. a) Location of strain gauges on the stud bolt,
b) Stresses in the stud bolt cross section of the steam generator primary collector during one year campaign.

to 350 MPa when the torque wrench was used. The measures
was accepted for the reduction of bending stresses in the
stud bolt. There ware an enlargement of the cover thickness, an elongation of stud bolts and use of the special
hydraulic device for prestressing of the stud bolts before
the tightening of the nuts. All parts of this joint obtain
the suitable position thanks to spherical surface of both
pads. The bending stresses decrease was theoretically to
50 MPa. Measurement during service proved the reduction of
bending stresses $\sigma_b = \sqrt{(\sigma_{bx}^2 + \sigma_{by}^2)}$ under 50 MPa, $\sigma_{bx} =$
$= 0,5(\sigma_{14} - \sigma_9)$, $\sigma_{by} = 0,5 (\sigma_6 - \sigma_{12})$, Fig. 5.

3 CONCLUSION

The owner of the power plant unit has the interest to extend its lifetime when design life of the components nears
to end. Extension is possible when the level of material
damage is known. The OFF-LINE diagnostic system is a good
tool for determination of this damage level. It uses measured data by sensors from I and C, DS and STM during service
conditions. Accuracy of results depends on perfection of
mathematical models of material damage processes. The
OFF-LINE diagnostic system can use these models, it is its
advantage.

REFERENCES

Lička, A. 1994. Design Service Condition of the Steam Gene-
    rator WWER 1000 MW for the NPP Temelin. Report of IAM Br-
    no, No. 1957/94 (in Czech).
Vejvoda, S., Moulis, R. and Vincour, D. 1996. Service Life
    Assessment of the Steam Generator WWER. *Proceedings of
    FAILURES' 96. Risk, Economy and Safety, Failure Minimisa-
    tion and Analysis*, pp 329 - 336. Edited by R. K. Penny.
    A. A. BALKEMA/Rotterdam/Brookfield/1996.

Ageing of Materials and Methods for the Assessment of Lifetimes of Engineering Plant, Penny (ed.)
© *1997 Balkema, Rotterdam, ISBN 90 5410 874 6*

Lessons to be learned from five unusual cases of high temperature material degradation

M.J.Molyneaux & J.B.Speck
Caltex Refinery, Milnerton, South Africa

ABSTRACT: Case studies of material degradation in service are presented as examples of failure mechanisms not commonly described in texts on mechanical or materials engineering. In all cases, the cause of failure was found to be due to a combination of long term exposure at high temperatures and transient service conditions, none of which were anticipated by the designers nor had been accommodated for in the long-term. Details of the most probable failure mechanisms are described in this paper, along with pragmatic solutions for the prevention or mitigation of these types of material degradation phenomena. A few lessons to be learned from these failures are pointed out within the context of equipment integrity over an extended life.

1 INTRODUCTION

Five unusual case studies of material degradation in service are presented as examples of failure mechanisms not commonly described in texts on mechanical or materials engineering. In all five cases, the cause of failure was suggested to be due to a combination of long term exposure at high temperatures and transient service conditions, none of which were anticipated by the designers nor had been accommodated for in the long-term.

Case 1 concerns oxidation and thermal fatigue cracking of Type 304 stainless steel refractory supports interacting with stress corrosion cracking. Case 2 concerns creep/fatigue cracking of a Type 321 stainless steel silencer initiated by stress corrosion cracking. Case 3 concerns thermal fatigue cracking of type 304H stainless steel cyclones in a Catalyst Regenerator Vessel propagated by oxidation product build-up. Case 4 concerns the creep failure of carburised alloy steel fired heater tubes. Case 5 concerns creep cracking of a low alloy steel reactor shell interacting with oxidation of the steel surface.

The specific combinations of high temperature service conditions together with either start up, shut down or otherwise peculiar operating conditions, give rise to mechanisms of failure which are not readily amenable to classical textbook models of failure.

The purpose of this paper is to outline details of suggested failure mechanisms and pragmatic solutions used for the prevention or mitigation of these types of material degradation phenomena.

2 CASE STUDIES

CASE 1: TYPE 304 STAINLESS STEEL REFRACTORY SUPPORTS

Numerous instances of failure by embrittlement and cracking have been observed in the liner plates, anchors and mesh for securing refractory to the shells of two Catalyst Regenerators on Fluid Catalytic Cracking Units where temperatures of around 700°C are typical. The liner plates, anchors and mesh were fabricated from type 304 stainless steel, a material which is commonly used as liner plates and as anchors or mesh inside high temperature vessels.

Under normal operating conditions, the oxygen concentration inside the vessel is around two percent with traces of sulphurous gases and therefore oxidation of type 304 stainless steel takes place only until a thin protective non-metallic layer forms over the surfaces.

Furthermore, any severe stresses induced in non-pressurised components fabricated from this type of steel due to thermal gradients and/or different thermal expansion coefficients between the austenitic steel and other ferritic steel components to which they have been welded can be expected to be relieved in the course of time as creep deformation takes place. So the choice of type 304 material at the design stage seems to have been a good one.

Nevertheless, embrittlement and cracking has been observed after as little as 20,000 hours service; this type of problem is prevalent in the petrochemical industry (Cantwell 1985). Obviously, sensitisation of the steel as well as a loss of ductility through precipitation of the so-called σ-phase is expected at operating temperatures of around 700°C, but in the case under discussion, ductility has been reduced to the point where the elongation after tensile testing to failure was as little as 1%.

Microscopic examination of samples of cracked and embrittled steel reveals an austenitic grain structure with smaller grains of the so-called σ-phase located typically at the triple points. Several forms of deterioration have been observed, namely, cracks and crevices, many of which are completely filled with a

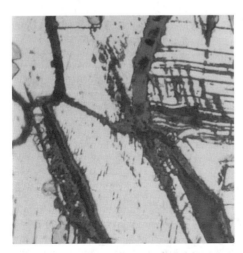

Figure 1. Microstructure (X500) showing oxide penetration into the grain boundaries

Figure 2. Microstructure (X50) intergranular crevices.

non-metallic phase (iron and chromium oxides and sulphates), penetration of the non-metallic phase into the grain boundaries of the surface regions of the steel (in some cases to a depth of two to three millimetres) as well as into those grains boundaries bordering on the cracks or crevices, transgranular and/or inter-granular cracks and carbide precipitates in all the grain boundaries.

Some penetration of the non-metallic oxide into the grain boundaries of those grains forming the surface regions may be expected to have taken place. This is common for is a common feature of alloys suffering from high temperature oxidation, but penetration to depths of two to three millimetres is uncommon after exposure to gascontaining 2% Oxygen and traces of Sulphur for as little as 20 000 hours at 700°C (Bolt & van Liere 1984).

This type of material degradation has been observed throughout the petrochemical industry (see Cantwell's survey results) and is believed to be caused by stress corrosion cracking when the vessels are opened, allowing air and moisture to come in contact with the sensitised and stressed components. Likewise moisture and air is introduced into the vessel in the early stages of plant start up. Either way, a corrosive, sulphurous acid forms on the oxidised steel surfaces which either cracks the stressed parts or preferentially corrodes the chromium depleted grain boundaries (details of this type of cracking mechanism are reported by Piehl 1964). As already mentioned, creep relaxation of severely stressed components at high temperature acts favourably while the vessel remains in service, producing significant though almost immeasurable changes in the shape and dimensions of the components concerned. However, these very changes are responsible for the reverse stresses experienced particularly in the surfaces of the steel when the vessel cools.

Stress corrosion cracking alone does not explain the extent of oxidation and embrittlement found in the surface layers of the steel. That can be explained as follows.

Once the surfaces of the steel have been cracked and damaged by corrosion, penetration of the non-metallic phase is promoted. Firstly, in those components where the surfaces exist in tension at ambient temperatures and where cracking occurs, pathways are provided by the cracks and crevices for movement of the oxidising gases when the vessel resumes normal service. Secondly, in those components where the surfaces exist in compression at ambient temperatures, preferential corrosion of the grain boundaries will be enhanced due to the lower activation energy for the corrosion reaction of the atoms in compression relative to the activation energy in the uncompressed state. Not only will inter-granular corrosion take place more readily, but the corrosion process will tend to reduce the stress level within the steel (Colombier & Hochmann 1967). On returning to high temperature conditions, the stress relieved structure will experience the same or similar tensile stress level as the previous cycle of operation. In the course of the new cycle however, instead of ordinary creep taking place, the corroded and thus embrittled surface will open up into micro fissures through which the oxidising gases move, allowing deeper penetration of the non-metallic phase into the steel.

One recommended practice for reducing the potential for embrittlement and cracking has been to wash all stainless steel surfaces with a soda ash solution when opening these vessels to atmosphere. Another solution is to ensure at the design stage that non-pressurised stainless steel components will not experience severe high temperature stress in service. This can often be achieved by changing the geometry of supports or liner plates, by providing expansion joints or slots in the mesh and liner plates and by applying additional refractory at particular locations for reducing thermal gradients in the steel to a minimum.

CASE 2: TYPE 321 STAINLESS STEEL SILENCER

A cylindrical stainless steel silencer assembly situated on the outlet pipe of a Catalyst Regenerator on a Fluid Catalytic Cracking Unit was found to contain several cracks originating in the heat affected zones of certain of the external welds. Cracking was first noticed after 140 000 hours service at temperatures varying between approximately 450°C and 650°C (depending on the cooling effect of weather and particularly wind conditions) and thereafter, the cracks were observed to have grown incrementally every time the unit was cooled to ambient temperatures. Occasional plant upsets sometimes result in temperatures estimated to reach about 750°C for durations of no more than 20 to 30 minutes.

The 8 metre tall silencer assembly conducts flue gases into the atmosphere via a 12 metre tall chimney stack and the internal gas pressure is only approximately 1 Bar. The only significant stresses in the 16 millimetre thick plates forming the 1.8 metre diameter shell are expected to derive from thermal gradients (around manways reinforcing plates and stiffening rings), from the weight of the chimney (which is butt-welded on top of the silencer), from any high speed winds, from residual stresses induced into the plate during the rolling operation to form the shell and from residual stresses induced in the welds and heat affected zones during fabrication. Other than the bending stresses due to wind, all of these stress components can be expected to be relieved in the course of time as creep deformation takes place. So high temperature degradation of the material at the design stage was not anticipated. Even the bending stresses due to wind loadings appear harmless in terms of the design calculations and ordinary fatigue considerations (ASME Vol. VIII Div. 1, 1968).

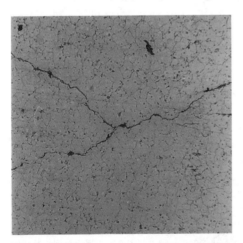 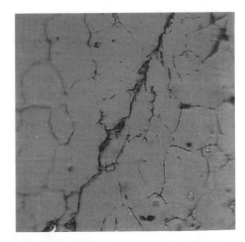

Figure 3. Intergranular cracking (100X)   Figure 4. Carbide precipitation (400X)

Microscopic examination using a portable optical microscope as well as examination of replicas taken from cracked portions of the steel surfaces revealed an austenitic grain structure with carbide precipitates both within the grains and prominently in all the grain boundaries. There was no evidence of the so-called σ-phase which is expected in austenitic steels exposed for long periods to temperatures above 650°C (Colombier & Hochmann 1967).

A clue to understanding the failure mechanism was the finding that the cracks were discovered only around manways and stiffening rings, only on one side of the shell, namely, the side from which the north-west wind comes in the winter season and only in the lower half of the silencer assembly where bending stresses due to wind loadings are expected to be the greatest. All evidence indicated therefore that crack propagation took place at high temperature in accordance with a creep/fatigue mechanism, conforming to one of the models described by Halford. Fluctuating bending stresses derive from wind loadings and are augmented by the tensile stress component produced by the differential cooling effect of wind on the stiffening rings and manways, more so on the wind side of the silencer than on the other.

The evidence appeared conclusive until official meteorology data was consulted. Historical charts showed that the prevailing wind is in fact south-east and that maximum windspeeds from the south-east are typically nearly three times those of winds from the north-west. So, wind alone could not be blamed for the material degradation.

Another clue was found in the discovery of small amounts of chloride on the oxidised external surfaces of the silencer. The chloride concentration of rain water was measured in the cold rainy season (when the north-west blows) and found to be

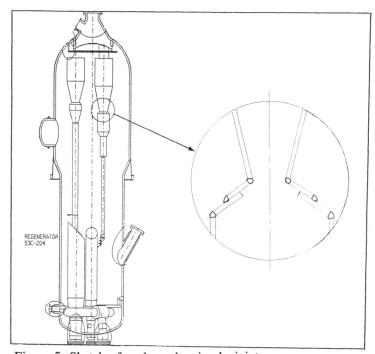

REGENERATOR
53C-204

Figure 5. Sketch of cyclone showing lapjoint

27

only around 6 ppm, but the rain water draining off the surfaces when the silencer is out of service was found to contain up to 65 ppm chlorides, clearly, the result of a concentration effect when raindrops evaporate on the hot north-west surfaces under normal service conditions. This concentration is sufficient to crack stressed stainless steel (Colombier & Hochmann 1967).

It was then concluded that small cracks were introduced into the surface as a result of chloride stress corrosion cracking during shutdowns in wet winter weather. The stresses are induced in the steel surface at low temperature as a result of high temperature creep deformation, as described in case 1. These surface cracks provide the stress intensification factor which accelerates the creep/fatigue process on the north-west side of the shell.

In the case under discussion, the problem of cracking was not solved, instead the stainless steel silencer was retired from service when the flue gas was re-routed to a new waste heat boiler. Nevertheless, this case underlines the importance of avoiding abrupt changes in profile (particularly at welds) and where possible features which may give rise to large temperature differentials in the design of austenitic steel structures intended for high temperature service. Where that is not possible or relevant (for example expansion bellows in high temperature flue gas ducts) then the structures should be provided with weather-proofing if service in coastal regions is intended and/or insulation should be applied to reduce thermal gradients.

CASE 3: TYPE 304H STAINLESS STEEL CYCLONE

A field weld joining the lower section of a cyclone to the upper section failed in service after nearly 150 000 hours operation at approximately 700°C. The type 304H stainless steel cyclone separates flue gas from catalyst inside a Catalyst Regenerator on a Fluid Catalytic Cracking Unit. The joint under discussion consisted of two 10mm thick plates overlapping by approximately 30 millimetres and fillet-welded on one side only to form a lapjoint.

Visual examination of the fracture face showed quite clearly that the weld had failed as a result of fatigue. Further examination and testing revealed that the twin cyclone at the opposite side of the vessel was partially cracked at the same field joint, though the fatigue crack had only propagated right through the weld at one point. All the other weld seams on the cyclones were then examined and found to be sound, except for a few fillet welds which attach reinforcing plates to the cyclone for structural supports. At certain locations the welds had cracked, allowing the plates to separate slightly from the cyclone at those locations.

Although possibly not the best of joint designs, in terms of both expected and actual operating conditions, the design of the field weld is quite adequate. If fatigue is to be blamed for the failure, then the explanation should involve significant stress variations. Several possible sources of stress variation were considered, for example, vibration resulting from the high speed gas velocities inside the cyclones. The most plausible explanation for this failure is a mechanism deriving from the accumulation of oxidation product between the two sets of plates joined by each of the fillet welds.

As already mentioned in case 1, the oxygen concentration inside the vessel is normally around 2 % with traces of sulphurous gases and therefore oxidation of type

304H stainless steel is only superficial. During start up however, oxygen concentrations are initially 20% after which they reduce to about 5% as heat is produced. Thereafter, while temperatures are maintained at 650°C and the flue gas pressure at 3 Bar, the oxygen concentration is estimated to reach about 10%. These conditions are only maintained for a few hours before normal operating conditions are achieved and traces of Sulphurous gases are introduced. During start up therefore, only superficial oxidation of type 304H stainless steel would be expected (Bolt & van Liere 1984). However, within the crevices formed between two sets of plates joined by welding, the high levels of Oxygen (possibly as high as 20%) would be maintained for much longer durations at the higher operating temperatures and diffusion of Sulphurous gases into the crevices would take place in due course. Within these crevices therefore severe oxidation of the mating steel surfaces would take place, limited eventually either by the availability of oxygen or by the availability of space between the plates for the accumulation of the volume of oxidation product. In fact, provided oxygen concentrations are sufficient and the oxidation product is sufficiently cracked or porous, the driving energy of the oxidation reaction will ensure that the reaction continues beyond the point at which the crevice becomes completely filled. The reaction would then stop only when the pressure build up between the plates reaches a particular ceiling determined by the activation energy for the reaction. Judging by the degree of bulging in the reinforcing plates and the amount of oxidation product observed in the crevices (up to 5 millimetres thick in some cases) a high order of magnitude of pressure is achieved.

Once these crevices become filled with oxides they act like wedges driving the plates apart. This situation is aggravated when the Regenerator cools down to ambient temperatures due to the difference in thermal expansion coefficient between the steel and the oxide. The combination of the geometry of the crevices and the oxidised condition of the steel surface at the bottom of the crevices constitute a high stress intensification factor for low cycle fatigue crack propagation which eventually gives over to high cycle fatigue failure once the stress levels in the remaining ligament of steel ahead of the crack approaches the yield point. In the case in question, failure resulted in the lower section of the cyclone falling to the bottom of the vessel.

Obviously, prevention of this type of failure in future requires any lapjoints to be designed with the minimum of plate overlap. On existing designs, any overlapping plates which form crevices should be perforated on one side to allow for the escape of oxygen which becomes trapped during start up.

CASE 4:   CREEP FAILURE OF 9%Cr/1%Mo ALLOY HEATER TUBES.

A Visbreaker heater processing vacuum distillate residuum oils by mild thermal cracking to gas, naphtha and fuel oil began to experience failures by through wall cracking of the 126 millimetre diameter tube coil after 160 000 hours service. This viscosity reduction simultaneously causes coke deposition on internal tube walls and an associated loss of heat transfer. The heater is periodically shut down after an operational period of about 120 days and is subsequently thermally decoked over 3 - 6 days at 670 - 695°C, prior to the next run. The internal pressures of steam and air during the decoke procedure remain below about 3 Bar.

The tube coil is fabricated in 9%Cr1%Mo material with an original nominal thickness of 7.0 mm, although between1 and 2 millimetres of steel has been lost in service mainly through oxidation and corrosion. In service metal temperatures range from 490 to 590°C and pressures from 14 to 28 Bar. In terms of ordinary creep or fatigue life calculations, a service life for the heater coil in excess of 300 000 hours would not be unrealistic. This is supported by the fact that apart from the material in the immediate vicinity of the cracks or at isolated areas where severe overheating had been experienced, no evidence of creep damage has been observed in the steel microstructure. The premature failure of the tubes was therefore unexpected.

Examinations of cracked tubes and others removed due to unacceptable loss in wall thickness, indicated that the tubes were experiencing a significant degree of carburisation of the inner surfaces, resulting in the appearance of a hard and brittle layer on these inside surfaces. Inner wall carburisation to depths of between 0.3 and 2.9 millimetres was measured. At isolated locations along the lengths of certain tubes, the embrittled carburised layers were found to contain extensive, randomly orientated cracks on the fireside of the tubes. The cracks were confined within the carburised layer except in those cases of tubes which leaked in service.

Optical microscopy carried out on samples of uncarburised tube material revealed a sparse presence of isolated cavities with partial grain deformation around the cracks, and a significant degree of fine carbide spheroidisation in a predominantly ferritic matrix. The average hardness of the material was found to be 194 HV. The tips of the cracks were found to be relatively blunted at the carburisation interface. In one instance, the uncarburised material ahead of a deep crack displayed deformation to such an extent that the crack had widened to approximately 2.0 mm.

The evidence suggests that carburisation had taken place at the high decoking temperatures of 670 - 695°C. Tube failure locations could not be related to specific regions in the heater or specific positions along the length of the tube coil, i.e. no correlation with respect to heat distribution in the firebox. Absolute tube metal temperatures rather than relative exposure time or tube age, appeared to be the most damaging feature of operational conditions in terms of carburisation depth. Thermal decoke periods only constitutes about one twentieth of the operating life of the coil - not unlike Perato.

A variety of explanations for crack initiation in the brittle material have been considered. For instance rapid cold pressurisation to 75 Bar on hydrotesting after repairs, operational variations like thermally induced localised stresses in areas where flame impingement has occasionally taken place, mechanical shocks due to hammering on the tubes at select locations to remove scale for thickness checks, or Austenitization of small patches of the carburised layer and subsequent re-transformation to Bainite at locations where over-heating took place beneath the last traces of coke to burn away during the decoking operation. Apart from the last one, the circumstances mentioned above apply equally to many other heaters on the same refinery without causing crack initiation on the inner surfaces of the tubes. Clearly, carburisation has been the key factor in explaining the failures.

Crack propagation appeared to be due to local stress rupture of the uncarburised material ahead of crack tips or by a mechanism of creep crack growth. The time scale for failure seems consistent with crack growth rates calculated using robust methods

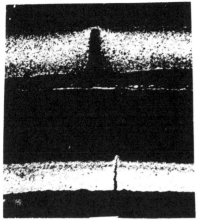

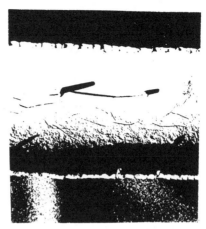

Figure 6.   Figure. 7

(Penny & Marriott 1995).

Although carburisation is a relatively common phenomenon in coke-producing heaters in the oil refining industry, the characteristics of our Visbreaker tube failures are peculiar and were unexpected. In view of the lost profit opportunities arising from the unscheduled outages a pragmatic solution for temporary repair was needed to restore the safe operability of the heater in the short-term.

A fitness for service assessment was pursued which was designed to: (a) burst all tubes containing undesirable carburised layer cracks by hydraulic testing; and (b) quantify remaining tube life using robust method calculations for mechanisms of stress rupture and creep crack growth.

The fitness for service assessment is the subject of a separate paper (Helmbold et al. 1997). However, it is worth noting that the goals of the assessment were successfully achieved. The Visbreaker heater was restored to an acceptable level of safe operability for approximately 9 months before complete coil replacement was undertaken with consequential significant loss savings.

Inner wall tube carbursation is primarily a product of the thermal decoking method employed in the Visbreaker heater under discussion. Several steps can be taken to prevent cracking of the carburised layer, such as a refinement of the decoking procedure to achieve better control over temperatures and prevent hot spots on the tubes. Alternatively, a mechanical method of decoking the tube coil (at much lower temperatures) could be considered. For practical purposes however, mechanical decoking methods would have to be accommodated at the design stage of the heater so that adequate access to tube internals can be provided.

CASE 5: LOW ALLOY STEEL REACTOR SHELL

After 140 000 hours service at a temperature of around 450°C and under an internal gas pressure of approximately 2.6 bar, the internal surface of a reactor was found to contain numerous shallow circumferential cracks. The cracks were located within a 15 millimetre circumferencial band immediately above a cylindrical shaped 100

31

millimetre wide doubling plate which is attached internally to the shell by fillet-welding around the upper edge and to which an internal semi-elliptical head is attached by butt-welding to the lower edge. Both the cylindrical doubling plate and the portion of the shell in question were rolled from 25 millimetre thick low alloy steel (11/4 Chrome 1/2 Molybdenum) and the internal head was forged from 32 millimetre thick plate of the same material. The longest crack measured 57 millimetres and surface grinding revealed that most were less than 2 millimetres deep, a few as much as 3 millimetres deep.

Microscopic examination using a portable optical microscope as well as examination of replicas taken from cracked portions of the steel surfaces showed that the cracks had all been completely oxidised and more closely resembled furrows filled with magnetite rather than cracks. The microstructure can be described as a fine grained ferritic matrix containing isolated grains of spheroidised pearlite with continuous carbides along the grain boundaries. Small quantities of creep voids were evident only around the extremities of the furrows.

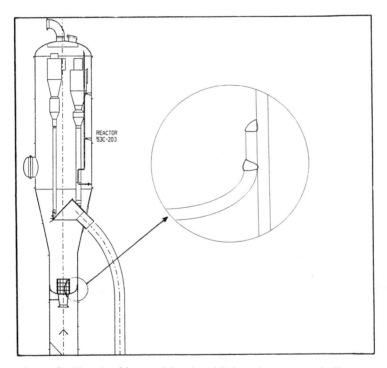

Figure 8. Sketch of internal head welded to the reactor shell

Considering the operating temperature of typically 450°C with occasional peaks of up to 500°C for no more than one hour at a time when the plant is started up or when plant upsets occur, some form of minor creep damage could be expected to have taken place at the exact locations in question. Bending of the shell due to temperature differentials in the shell around the doubling plate and due to the abrupt transition in effective shell thickness at the location of the fillet weld would result in tensile stress precisely where the cracks were found. However, any creep deformation would result in stress relief and therefore creep damage, if any, would be self-limiting and so

32

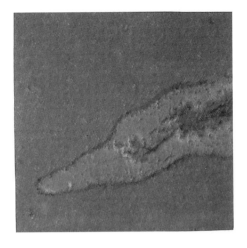 

Figure 9. Replica of oxidised crack tip (X100)   Figure 10. Same replica (X200)

slight that it would be difficult to detect through macro- and ordinary micro-examination.

Cracking purely as a result of cyclic loading is also unlikely: Temperature fluctuations have been measured during normal operation and these are typically less than ten degrees so stress fluctuations due to thermal cycling would be insignificant. Furthermore, the unit had experienced less than thirty start-ups or upsets when the cracks were discovered, so crack initiation through low cycle creep/fatigue (see for example Campbell) can be precluded. At temperatures of around 450°C, the very high order tensile stress required for crack initiation (within thirty cycles) cannot be maintained because the long durations between start ups would ensure sufficient stress relief and result in the amplitudes of subsequent stress cycles being reduced to insignificant magnitudes.

The internal head design is quite acceptable in terms of the relevant code of construction (ASME Vol. VIII Div. 1, 1968) and yet the early stages of failure are evident. Clearly, in terms of the classical failure mechanisms and mathematical models of high temperature material degradation, the existence of these cracks cannot be reasonably explained.

A clue to a reasonable explanation was found in the existence of a hard brittle magnetite layer which forms on the steel surfaces during plant start ups when a high temperature oxidising atmosphere prevails inside the reactor for periods of 12 to 24 hours. Thereafter the atmosphere becomes reducing and remains so until the next start up event. As temperatures rise, the magnetite layer thickens, the degree of bending and tensile stress levels in the shell at the failure site magnify and the chances of the layer cracking at the point of highest strain increase. If the layer

cracks, the exposed steel at the bottom of the crack oxidises further and this introduces a magnetite filled notch into the steel surface The notch would reach its greatest size at the highest temperature achieved during plant start up. Subsequent cooling to the normal operating temperature would result in an additional tensile stress component being induced in the steel at the deepest point of the notch due to the large difference in thermal expansion coefficients between steel and magnetite. The stress concentration around the notch would then produce localised creep damage at the deepest point, resulting in incremental propagation of the notch deeper into the steel. During each start up event, the scenario describe above would repeat, with the crevice providing the preferred site for cracking of the magnetite layer on each cycle.

The mechanism of rupture, crack formation and propagation proposed above has been arrived at by a process of deduction, guided by the discussion of this subject by Penny & Marriot 1995). Although speculative, the proposal is nevertheless consistent with the finding that the crevices were all completely filled with magnetite, the finding that voids were evident only at the edges of the furrows and the finding that micro cracks or fissures were not evident in any of the numerous replicas taken from the different damage sites.

The potential for cracking in future was reduced by depositing additional weld metal at the location in question to produce a more gradual transition in effective shell thickness at the fillet weld. Subsequent inspection after a further 24 000 hours service revealed no further evidence of cracking. Another possible solution would be to weld overlay the susceptible areas using a suitable Nickel based alloy which offers better oxidation resistance than low alloy steel.

3. LESSONS TO BE LEARNED

A lesson to be learned from the cases presented in this paper is that designers do not always have (at the design stage) all the relevant information on service conditions to ensure equipment integrity for an extended life. This is where feed-back from the end users (plant operators) on failure mechanisms is of great importance to designers for improving their designs or revising Operating Manuals.

These case studies also serve as a warning to plant operators and experts on remnant life assessment that even when all the usual failure mechanisms have been considered and studied through degradation models and formulas for remnant life predictions, there is still the possibility that a component could fail prematurely by a hazardous, unexpected combination of mechanisms, because few authors (notable recent exceptions being (Chalenger et al,) (Dyson & Osgerby 1988) and (Penny & Marriot 1995)) have warned about such hazards.

When remnant life assessments are required for critical items of equipment, the cardinal lesson to be learned is the importance of performing a re-examination of the equipment and a re-assessment of the life expectancy at well-planned intervals, in order to ensure the accuracy of the remnant life predictions. Much depends on reliable methods for defect detection as well as robust methods which can take into acount sparse materials data usually available for aged components.

REFERENCES

ASME, Pressure Vessel Design Code, Vol. VIII, Div. 1, 1968.

Bolt, N., van Liere, J. Fireside corrosion of stainless steel SA213TP 347H tubes in reheaters and superheaters of oil-fired power generating units. *Stainless Steels `84*, pp 554-561, The Institute of Metals, London.

Campbell, R.D. (1971) Creep-Fatigue Correlation for 304 Stainless Steel Subjected to Strain Controlled Cycling with Hold Times as Peak Strain, *ASME Paper No. 71-PVP-6*

Cantwell, J.E. (1985) Embrittlement and Intergranular Stress Corrosion Cracking of

Stainless Steels after Elevated Temperature Exposure in Refinery Process Units, *Recent Metal Failures*, Caltex Petroleum Corporation, Dallas, Texas.

Colombier, L., Hochmann, J. (1967) Stainless and Heat Resisting Steels. *Edward Arnold (Publishers) Ltd*, Great Britain.

Dyson, B.F. and Osgerby, S. (1988). Modelling Synergy between Creep and Corrosion for Engineering Design, in *Materials '88, Proceedings Inst. Metals Conference on Materials and Engineering Design*, London.

Halford, G.R. (1967) Evolution of Creep-Fatigue Life Prediction Models, *ASME Special Publication AD*, 21.

Helmbold, A.C. et al & Penny, R.K., 1996. *Fitness for Service Assessment of a Damaged Heater Coil*, Proc. FAILURES '96, Rotterdam: A.A. Balkema Publishers.

Penny, R.K. & Marriott, D.L., (1995) Design for Creep - second edition. *Chapman & Hall*, London.

Piehl, R.L. (1964) Stress Corrosion Cracking by Sulfur Acids. *29th Midyear Meeting of the American Petroleum Institute's Division of Refining*, St. Louis, 11 May 1964.

Life assessment techniques

Structural integrity assessment and lifetime prediction

B. Tomkins & N. M. Irvine
AEA Technology, Warrington, UK

ABSTRACT: For major components in engineering plant and infrastructure lifetime prediction and management are key issues. The design basis with associated quality control on fabrication and construction provides the original fitness-for-purpose justification and can be used for lifetime judgements for a considerable time in operation. However, as design life is exhausted or operational conditions change reassessment is necessary and this invariable involves a current structural integrity evaluation. It is the quantification of integrity which is different and this paper will examine the methods which have been developed to make such assessments for ageing plant. This involves both an interdisciplinary technical approach and one which is economically viable. Examples will be given to illustrate the developing technology.

Ageing of Materials and Methods for the Assessment of Lifetimes of Engineering Plant, Penny (ed.)
© *1997 Balkema, Rotterdam, ISBN 90 5410 874 6*

Decision-making for boilers with long-time operation

C. Delamarian
Institute of Material Testing and Welding, Timisoara, Romania

H. R. Kautz
Grosskraftwerk Mannheim AG, Germany

ABSTRACT: Utilities in many countries around the world attempt to extend the service time of their plants. The operating time exceeds 100,000 hours, 200,000 hours or even more. This will raise problems which will increase in the future because of economical and political reasons that enforce further exploitation of this old equipment. A big question mark is the further safe operation of these boilers. Is there any possibility to maintain such old boilers in operation? If there is one, which facts support such a decision? People charged with the maintenance-related problems of these old boilers are facing an enormous responsibility. The purpose of this paper is to offer some answers to these questions and also a *solid* background for the decision making with respect to boilers with longtime operation. Some interesting aspects related to design, component condition evaluation, life assessment, and life extension of old power plant components will be briefly presented. Mathematical as well as practical approaches are considered in order to provide an insight into damage processes affecting power plant components with longtime service.

1 BACKGROUND

During the last decades, the construction of new conventional power plants has come to a virtual halt in many countries around the world. Some of the factors responsible for this situation are:

- Industry's perception of the existence of sufficient reserve capacities;

- uncertainty in forecasting the increase of the energy demand;

- high investment costs for new plants;

- applicability of more stringent environmental standards for new plants.

In the U.S.A., by the year 2000, it is estimated that nearly 20 % of all the fossil-fired units will be more than 40 years old, and almost 44 % will be more than 30 years old. Based on common understanding, nearly half of all fossil-fired plants have been retired before their fortieth year and three quarters before serving 50 years. In the future, this is unlikely to continue because of the shrinking reserve capacities and the long lead time for the addition of new capacities [1].

In Japan, in 1989, more than 60 % of the thermal power plants were classified as aging plants with operating hours exceeding the designed 100,000 hours [2]. In Ukraine there are boilers with a service time up to 300,000 hours [3].

Statistics show that presently 50 to 70 % of the German power plants exceed their design life of approx. 20 years. Towards the end of the century, 15 to 25 % of the power plants will have reached the 40 years limit and have to be repaired according to condition in order to mitigate or eliminate the old design-induced or operation-induced mistakes to keep the systems further in service [4].

In Italy, 126 units with a total capacity of about 31,000 MW have an average operating time exceeding 100,000 hours with a few plants with even more than 200,000 hours [5].

A similar case may be found at the Mannheim Central Power Plant (Grosskraftwerk Mannheim - GKM), Germany, where one boiler approaches and another one exceeds 200,000 operating hours.

A big question mark is the continued safe operation of these boilers. Is there any possibility to maintain in operation such old boilers? If there is one, which facts support such a decision? The purpose of this paper is to offer some pertinent answers and also a *solid* background for the decision making with respect to boilers with longtime operation.

2 DESIGN CALCULATION AND SAFETY FACTORS

The design of industrial components and structures includes safety aspects such as:

$\Rightarrow$ Safety factors: The value of every factor depends on the type and purpose of the structure. In power plant design, a safety factor of 1.5 is usually used for components that are exposed to creep and thermal fatigue during plant operation (there are components that are designed with a safety factor of 2 or 3). That means the values of the material characteristics used in designing boiler components (R_m, $R_{P0.2}$, creep strength, fatigue strength etc.) represent 0.666 maximum of the real material characteristic value provided in the manufacturer's material quality certificates.

Starting in spring 1953, the DVM[*] creep limit (pursuant to DIN Standard 50 117) was abandoned as calculation basis. It was substituted by

the 100,000 hours strength (mean value) determined in a long-time test. In order to obtain adequate safety against rupture within 100,000 hours a simple safety factor ($\sigma_{B/100,000}$) of 1.5 was used. A recalculation should be made to check, whether the 1 %-100,000 hours elongation limit ($\sigma_{1/100,000}$) was not exceeded and whether with a temperature excursion by 15 K there is still the 1.0 times

---

[*] German Association of Materials Testing

safety against $\sigma_{B/100,000}$ exists (temperature margin)[*] ;

$\Rightarrow$ a safety margin of the operating pressure is used in the component wall thickness calculation. A maximum pressure value shall consider the worst case that may occur during boiler operation because of unexpected over-loads;

$\Rightarrow$ a safety tolerance for the tube wall thickness shall be selected in order to compensate for the loss in thickness due to the corrosion erosion over the design operating time;

$\Rightarrow$ for the manufacture of these components a standard wall thickness is selected equal to or one magnitude thicker than the design calculated thickness.

So in component design the design tube wall thickness will increase. Therefore, a smaller real stress in the tube wall cross section will occur during boiler service (compared with the one corresponding to the calculated minimum wall thickness with respect to the operating load) induced by the operating pressure. Under these circumstances, the creep damage rate illustrated in a diagram - strain versus time - for a constant temperature value will have another pattern (*Fig. 1*) and therefore, a real safe operating time may be expected. In another diagram (*Fig. 2*), where lg(sigma) is plotted versus lg(time) at a specific temperature the real safe operation may be compared with the design operation. The magnitude of the ratio t_{real}/t_{design} (Fig. 2) may vary from one component to the other, from one boiler to the other being a function of many factors such as:

- value of the safety factor used in the design calculation;

- dimensions of the component (outer diameter, wall thickness);

- safety margin selected versus the corrosion rate that occurs during plant operation;

- other damage mechanisms affecting the plant component;

- quality of the operating medium (water, steam, flue gas) coming in direct contact with the component surface;

operating conditions such as temperature, pressure, load system etc.

Formerly, in power generation the lack of knowledge of the failure mechanisms affecting plant components was frequently compensated by increasing the wall thickness of plant components in order to ensure a safe plant operation over the design life. In the last

---

[*] **New design rules**: in 1967, in order to avoid a simple wall thickness increase due to a reduction of the creep strength values, the components were designed for a specified minimum service life. The utilities considered 200,000 hours minimum service life, adequate to a full load operation of 8000 hours annually, to be an adequate value. Starting 1969 the minimum creep strength up to 200,00 operating hours is used as design calculation value. So instead of the mean values the values of the lower limit curve of the the creep strength scatter (-20 %) must be used.

decades, intense investigations were conducted to study the real component behavior and the real material condition during plant operation. A more accurate knowledge of these factors allowed plant operators to take proper measures in order to reduce, minimize or prevent the damage.

The conclusion is that old components may achieve longer operating times beyond the design life sometimes mainly because of their inaccurate (too conservative) life calculation.

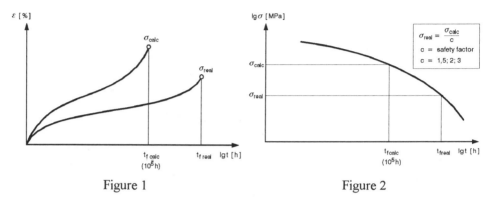

Figure 1   Figure 2

3   PROBABILISTIC APPROACH IN LIFE ASSESSMENT OF POWER PLANT COMPONENTS

In case of engineering structures where components are exposed to different damaging processes during operation, the application of a probabilistic approach in the prediction of the component time-dependent behavior is a common method. Such an approach allows to assess the probability that a specific event (e.g. the failure) will occur during the operating process. Failure probability values in the range of 10^{-6} to 10^{-12} are usually used in the design of some engineering structures, and they depend on the following factors:

a)   plant value;

b)   cost-related consequences of a possible failure;

c)   failure consequences for the human safety;

d)   environmental impact of a possible failure etc..

In the field of energy generation, time-dependent damage processes occur. Creep, fatigue, thermal fatigue, corrosion, erosion, etc. are such processes. That means there is a gradual increase in the failure probability over the boiler service time. Usually a failure probability value in the range of 10^{-6} - 10^{-8} is selected for the design calculation of power plant pressure vessels.

A probabilistic approach to the remaining life calculation of power plant components may be used to assess the component life time, the time when an inspection, repair, replace is required [6]. This mathematical method is applied also in the design for which mean values of material characteristics obtained from specific material tests are selected and in the crack growth rate calculations or in bulk creep damage assessment. The time to failure may be correlated with a specific failure probability value (e.g. 10^{-3}). Even in the case, when the failure probability level increases above a specified value, (*Fig 3a* for creep and *Fig. 3b* for low cycling fatigue - Wöhler diagram) the component may continue to operate safely, but measures are required to reassess the remaining life.

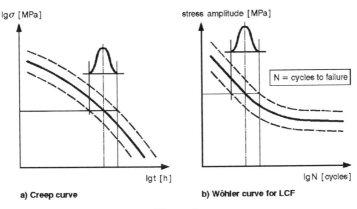

a) Creep curve   b) Wöhler curve for LCF

Figure 3

There are actually many non-destructive examination procedures to estimate the remaining life of power plant components exposed to creep or fatigue [8, 9]. A detailed discussion will be presented below.

Other alternatives of approaching the increase of the failure probability are to extend the component life by repair or to reduce operating parameters [10]. The effect of applying such methods with respect to the failure probability is illustrated in *Fig. 4.a* and *4.b*. Some alternatives of rehabilitating old components will be described below.

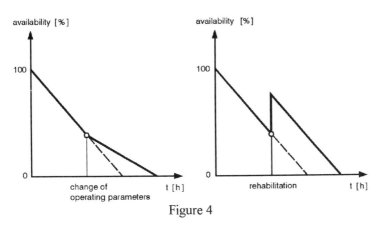

Figure 4

4   SERVICE LIFE ASSESSMENT OF CREEP-FATIGUE DAMAGED COMPONENTS

One of the most recent methods of power plant maintenance is the condition-oriented maintenance [7]. Here a detailed examination of the component condition is performed in order to assess the real time of safe operation. Being aware that the creep damage may occur in a bulk (cavitation) or localized (cracks) form, the component condition may be assessed after performing surface crack examinations (liquid penetrants, magnetic particles, ultrasonic examination, etc.) or by means of replication in critical locations. According to the specifications of some European guidelines such as

⇒ NORDTEST 107/1992 „Reference Micrographs for Evaluation of Creep Damage in Replica Inspection"

⇒ VGB-TW 507/1992 "Guideline for the Assessment of Microstructure and Damage Development of Creep Exposed Materials for Pipes and Boiler Components",

the creep damage (by cavitation or by microstructure degradation) may be assessed and classified in five assessment classes. For any class recommended actions or required inspection intervals are provided within several European and U.S. guidelines such as:

⇒ VGB Guideline 507 [8];

⇒ APTECH/EPRI (*Fig. 5*);

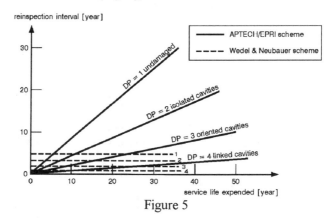

Figure 5

⇒ Manual for Maintenance and Retrofit of Conventional Power Plants [9].

If bulk creep damage is observed in a critical component location, recurrent examinations are required in order to monitor the damage evolution. For instance:

• If a creep damage of Assessment Class 2 (according to VGB-TW 507/1992 or NORDTEST 107/1992) is found, a recurrent inspection is required within 20,000 operating hours [11];

• if creep damage of Assessment Class 3 (according to the same source) is diagnosed, a recurrent inspection is required within 15,000 operating hours [11];

The damage level may be differently evaluated with respect to the respective operating time of the component. For instance, when Assessment Classes 2 or 3 are detected

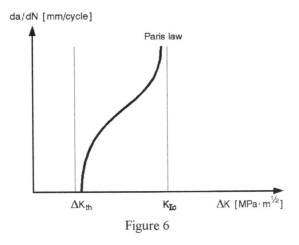

Figure 6

after 100,000 operating hours, the time for the next recurrent inspection shall be doubled (i.e. after 40,000 and 30,000 operating hours respectively) [11]. Another approach is illustrated in *Fig. 6* [1].

Based on Japanese statistics more than 90 % of the problems related to power plants occurred in boiler tubes. The causes are 47 % low cycle fatigue, 11 % creep and 16 % corrosion and wear [2]. From the statistics, the criticality of low-cycle fatigue (LCF) damage in power plants may be derived. LCF usually materializes in the form of fatigue cracks. Many mathematical models were developed in order to assess the LCF crack growth. For the calculation of the crack growth rate versus the stress intensity factor the Paris law and the corresponding diagram are used (Fig.6). For the assessment of the life time number of load cycles) of a component exposed to LCF versus the stress range the Wöhler diagram is used (see Fig.3b).

If localized creep (or fatigue) damage is observed (cracks), a crack growth calculation shall be performed in order to assess the crack growth rate and the time when the crack will reach a critical size (depth or length), and actions must be taken. Saxena and coworkers [12-14] suggested that the mean time-dependent crack growth rate, $(da/dt)_{avg}$, can be correlated with the mean value of C_t, $(C_t)_{avg}$ parameter during the hold time (constant temperature and stress) when creep sets in:

$$(da/dt)_{avg} = \frac{1}{t_h} \cdot \left[da/dN - (da/dN)_0 \right] \qquad (1)$$

and

$$(C_t)_{avg} = \frac{1}{t_h} \cdot \int_0^{t_h} C_t \, dt \qquad (2)$$

wherein

a   - crack depth
t   - time
N   - number of cycles
da/dt   - crack growth rate

47

If the component operation is considered no longer safe because of the crack size, a repair shall be considered. In any case, the retirement of a component affected by bulk or localized creep damage or localized fatigue is no longer considered an ultimate technical solution.

5 LIFE EXTENSION OF OLD COMPONENTS

New welding techniques developed in the last decade support the maintenance process of old plants. Once the accuracy of non-destructive examinations reached very high levels, the detection of damaged locations became more and more precise. Cracks as a result of creep or fatigue damage may be successfully repaired, in some particular circumstances by way of welding. Welding techniques that use Ni-based alloys as a filler were successfully used to repair old components manufactured of ferritic materials without performing post-welding heat treatment (so-called „cold welding technique" [15]). The major advantage of this procedure is the low level of welding-induced stresses. Nevertheless, full attention shall be given to the evaluation of the local material damage. This procedure shall not be applied to those components that are strongly affected by different damage processes. The decision to repair a damaged component shall be made based on an accurate component examination, on economic considerations, on the expertise of the maintenance personnel and shall be tailored to each case.

Another welding procedure that gained a wide application in the power generation industry (not only in conventional power plants, but also in waste-to-energy plants or nuclear plants) is the weld metal overlay technique. The principle of this procedure is illustrated in *Fig. 7* [16]. By way of this procedure old boiler components (waterwall tubes, superheater/reheater tubes) strongly affected by corrosion damage may be rehabilitated. The welded overlay will increase the corrosion resistance of the external tube surface and will also restore its initial wall thickness. Full attention shall be paid to the welding procedure including weld material selection and weld parameters. Such a procedure may extend the component life by ten years in accordance with the manufacturer's guarantee [16].

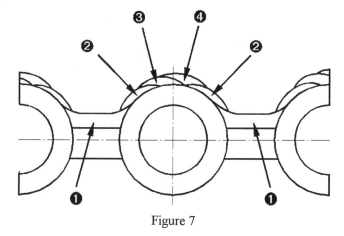

Figure 7

CONCLUSION

A plant with an operating time approaching or exceeding 200,000 hours actually represents a technical challenge. Solutions are available today in supporting the decision making of the maintenance of such plants. The availability of power plant components was increased over the past decades by way of using new maintenance strategies. These strategies are mainly oriented on the more accurate understanding of the component material condition and damage processes or mechanisms affecting plant components and also on reducing or eliminating the causes of such processes. New welding techniques developed in the last decade were successfully applied to the rehabilitation of damaged components. New lessons were learned from recurrent mistakes in every department of the energy generation industry beginning with ordering over design, construction, manufacturing to operation. The retrofit of older plants may mitigate or eliminate old design-induced or operation-induced mistakes to keep the system further in service.

The conclusion that may be reached from this paper is that a boiler with 200,000 or even 300,000 operating hours must be more carefully treated with respect to the maintenance, but should not be retired from service. Adequate safety margins allow a continued operation of these old boilers.

REFERENCES

[1]   Viswanathan, R.: „Life Assessment of High-Temperature Components - Current Concerns and Research in the U.S.“; Proceedings of the European Conference „Life Assessment of Industrial Components and Structures“, Churchill College, Cambridge, UK, 1993.

[2]   Nakanishi et. al.: "Life Prediction Assessment and Reliability Technology for Boiler Plants", Thermal and Nuclear Power, Vol. 40, p 1104 (1989) quoted in the paper of Ogata, T. and Nitta, A.: „Japanese Activities in the Area of Power Plant Life Extension“; Proceedings of the European Conference „Life Assessment of Industrial Components and Structures“, Churchill College, Cambridge, UK, 1993.

[3]   Internationale Technisch-Wissenschaftliche Konferenz „Schweissen in der Energietechnik“ (welding in the power industry - in German), Kiev, Ukraine, 1996.

[4]   Kautz, H.R.: „Management of recurrent mistakes in power industry SP249-end users response“, in „Failures ‘96“ Proceedings of the Second International Symposium on Risk, Economy, and Safety, Failure Minimization and Analysis“, Pilanesberg, South Africa, 22-26 July 1996.

[5]   Billi, B.; Regis, V. and D'Angelo, D.: „Third Generation Structural Integrity Assessment Methodologies for Residual Life Evaluation of Thermal Power Plant Components“; Proceedings of the European Conference „Life Assessment of Industrial Components and Structures“, Churchill College, Cambridge, UK, 1993.

[6]   SPRINT 249 Project „Implementation of Power Plant Component Life Assessment Methodology Using a Knowledge-Based System“

[7]   Kautz, R.: „Zustandsorientierte Instandhaltung“, VGB Kraftwerkstechnik (71) 1991, S. 653-657 (Condition-Based Maintenance)

[8]   VGB-Richtlinie R 509 L. VGB - Technische Vereinigung des Grosskraftwerks-
      betreiber e.V., Essen, 1988" (Recurrent Examinations - Guideline)

[9]   Manual for Maintenance and Retrofit of Conventional Power Plants, ISIM-GKM
      Project, 1996.

[10] Dehelean, D., Delamarian. C., Flueras, D.: „Nachrüstung von Komponenten
      rumänischer Kraftwerke mit Hilfe der Schweißtechnik" (Retrofitting of Romanina
      power plant components by welding - in German), DVS Konferenz, Hannover,
      Deutschland, Sept. 1996, Conf.Proc. p.167-174.

[11] IVO Recommendations, IVO Finland, 1992.

[12] Saxena, A. and Gieseke, B.: „Transients in Elevated Temperature Crack Growth",
      Proceedings of MECAMAT - International Seminar on High Temperature Fracture
      Mechanism and Mechanics, vol. 3, 1987, pp. 19-36.

[13] Gieseke, B. and Saxena, A.: Correlation of Creep-Fatigue Crack Growth Rates
      Using Crack-Tip Parameter", Advances in Fracture Research, ed. K. Salama et al,
      (Elmsford, NY: Pergamon Press, 1989), pp 189-196.

[14] Yoon, K.B., Saxena, A., and McDowell, D.L.: „Influence of Crack Tip Cyclic
      Plasticity on Creep Fatigue Crack Growth", 22nd ASTM National Symposium on
      Fracture Mechanics (Philadelphia, PA: ASTM, in press).

[15] Allen, D.J. et al: Cold Weld Repair - Development and Application. Welding and
      Repair Technology for Power Plants, EPRI Conference at Daytona Beach, FL,
      1996

[16] French, S. and Rumbaugh, K.: Fireside corrosion erosion mitigation via the appli-
      cation of weld metal overlay. Welding and Repair Technology for Power Plants,
      EPRI Conference at Daytona Beach, FL, 1996

Assessment of a transmission gas pipeline destruction

F. Valenta, J. Michalec, S. Konvičková, M. Růžička, J. Řezníček, M. Sochor
& M. Španiel
*Czech Technical University, Faculty of Mechanical Engineering, Department of Elasticity
and Strength of Material, Prague, Czech Republic*

ABSTRACT: An original approach to the assessment of the onset cause of a DN900
transmission gas pipeline breakdown was applied by the research team of the Czech
Technical University (CTU). Using retrieved parts of the damaged pipes an extensive
surface planary corrosion defect (where the rupture initiated) was reconstructed. Taking
an intact pipe from the breakdown vicinity, an identical fault was machined on its surface
and such a prepared testing body was equipped (while using large deformation strain
gauges) for a hydrostatic burst test. Based on the authors' original FEM software and
methodology, the strain gauge placements and the limit pressure prediction were assessed,
after which being verified by the experimental data. ANSI/ASME standards were applied
as well.

1 INTRODUCTION

Submitted to examination was the problem of a rupture on a DN900 transmission pipeline,
having nominal thickness of 12 mm, made of St 52.3 steel, spirally welded pipes. The
pipeline was supposedly in a convenient operational state with its inspection planned for
the next year, when, unexpectedly, an explosion (at the maximum working pressure not
exceeding 5.6 MPa) destroyed its part. The breakdown assessment complicacy consisted
in determining the basic cause of the onset of the extensive fracture comprised of a number
of pieces of the broken and thoroughly deformed pipeline part which having the length of
about 22 m (Figure 1(a)).

After all parts of the damaged pipes having been picked up, it was stated that the
rupture had started in an extensive surface planary corrosion defect (Figure 1(b)). The
operator and university workers decided that the defect dimensions be reconstructed,
from which a model planary defect be designed serving for both a computational and
experimental approach to the problem. During the breakdown, the corroded pipeline
part had undergone an intensive plastic deformation, and, subsequently, a considerable
thickness reduction. Thus, it is evident that there, above all, the main difficulties were to
be encountered when the model defect thickness distributions were to be assessed (which
were to correspond most closely to the actual defect state prior to the rupture). For this
purpose, utilized were the CTU research results of the limit analyses of transmission gas
pipelines having corrosion defects, which had previously been carried out on the DN800
- X60 pipelines; i.e. (based on foregoing similar hydrostatic tests) areal estimation of the
remaining pipe thickness on the fracture line was carried out.

With respect to the complexity of the defect topography, it was decided that large deformation strain gauges be placed in significant localities where both positive and negative extremes should appear (according to preliminary FEM computations) and to thus confirm their quality. In order to make more accurate the conclusions on the breakdown cause, further computations were worked out using a finer FEM mesh and applying alternative defect remaining thicknesses of 4 and 3 mm, respectively. Likewise, ANSI/ASME standards were used (in all three known modifications). A complex of all results of the presented theoretical and experimental analyses enabled estimating the breakdown onset conditions of the DN900 pipeline.

Figure 1: General view of the broken-down gas pipeline fragments(a) and retrieved parts of the damaged pipe where the responsible corrosion defect can be seen(b).

2 MODEL BODY DESCRIPTION AND EXPERIMENT

The model surface defect (to be utilized for the pipeline breakdown assessment) was designed on the basis of investigated actual defect geometry, remaining thicknesses and material characteristics of the ruptured DN900 pipeline: the model defect contour coincides with the actual defect contour; the model defect wall thickness was proposed (according to the ruptured pipeline actual wall thickness) to be changed from 12.4 mm to 4 mm; issuing from the data of the material ductility, and the wall thickness topology on the fracture line, the minimum model defect wall thickness was proposed to be 4 mm. Figure 2 shows the model defect contour across which dashed lines mark positions of six planary sections: one in longitudinal and five in circumferential directions, the profiles of which are shown in Figure 3. In the middle of the defect contour, a field was machined whereof having the thinnest (4mm) pipe wall in the defect.

The model surface defect was machined on a model body consisting of an intact pipe taken from the accident vicinity, and, subsequently, equipped for strain measurement. After the preliminary FEM computation determining, theoretically, the strain extremes, the strain gauge crosses were placed onto these localities (Figure 4). The experiment consisted in hydrostatic tests including preliminary testing cycles and the burst test proper which ended in the model body rupture at a pressure of 8.1 MPa. In the course of these tests, the model defect strain distributions were recorded using the strain gauges for large strain measurements, the plots of which are shown in the conclusion.

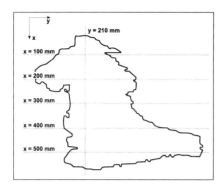

Figure 2: Contour of model defect with sectional positions.

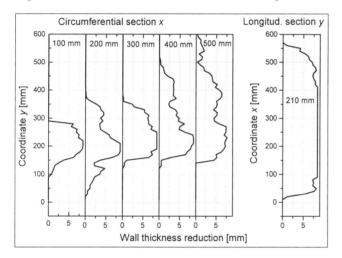

Figure 3: Model defect sections - see Figure 2

3   LIMIT ANALYSIS

This presented limit analysis is based on the CTU research team results and findings, parts of which were delivered in turn at the CAPE'93 and CAPE'95 colloquia (Valenta, Sochor, Španiel, and Michalec 1994), (Valenta, Sochor, Španiel, Michalec, Růžička, and Halamka 1996). Its idea consists in finding such a pressure at which a relative plastic area length of the defect reaches a certain limit value:

$$\Lambda = \frac{L_P}{L_C} = \Lambda_{LIM} \tag{1}$$

where

L_P ... representing a found plastic area length depending on a strain limit value ε_{min} determined from experiments (see Figure 5)

$L_C = L_D - 2 \cdot L_W$ ... representing the central part of the defect length (called "defect core") defined as an axial distance between two nearest displacement distribution wave hollows (called "defect length" L_D) reduced by two "wave lengths" (L_W) of an ideally symmetrical elastic pipe (see Figure 6)

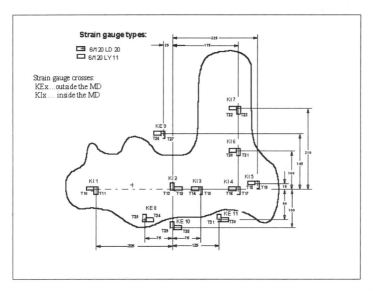

Figure 4: Placement of strain gauge crosses in/around the model defect (MD).

The limit analysis demonstrated in (Valenta, Sochor, Španiel, and Michalec 1994), (Valenta, Sochor, Španiel, Michalec, Růžička, and Halamka 1996) had been developed for DN800 pipes made of material X60 (with very satisfactory results), and thus, not adhering to the quite proper input presumptions, the limit analysis methodology – applied on DN900 pipe made of the St52 steel (being markedly stiffer) – was considered as to be an informative one only. In order to cope with a certain inconsistency between the limit analysis rate required in the preliminary phase (when, in a short time, the model defect minimum thickness $t_{min} = 4$ mm , estimated from the broken-down pipeline fragments, was to be confirmed; and the strain gauge positions in the model defect were to be determined) and the accuracy and reliability requirements being put on the final breakdown cause assessment, used were two FEM basic models, based on the minimum thickness area in the defect having $t_{min} = 4$ mm , with different mesh quality. The two model meshes were marked as CM4 for the coarse mesh and FM4 for the fine mesh and differ as follows: CM4/FM4 having 448/1489 isoparametric elements, with the number of nodes and degrees of freedom being 2160/7728 and 6480/23184, respectively.

Radial displacement distributions (RDD) of the model defect (based on the FM4 FEM computational model) were computed for the line given by the strain gauge crosses of KI1–KI4. The RDD is plotted in Figure 6 only for $p = 10.5$ MPa . Plastic regions were also investigated for these pressures (as an example, shown in Figure 5 is a length of plastic region being $L_P = 575$ mm and corresponding to $p = 10.5$ MPa , $\varepsilon_{LIM} = 0.03$). For the $p = 10.5$ MPa , measured was the defect length $L_D = 880$ mm and computed was the wave length corresponding to the rotationally symmetric elastic bending of the DN900 pipe margin to be $L_W = 181.5$ mm . From this (see Figure 6), the defect core follows as

$$L_C = L_D - 2 \cdot L_W = 537 \text{ mm} \tag{2}$$

and the relative plastic area length of the defect being

$$\Lambda = \frac{L_P}{L_C} = \frac{575}{537} = 1.07. \tag{3}$$

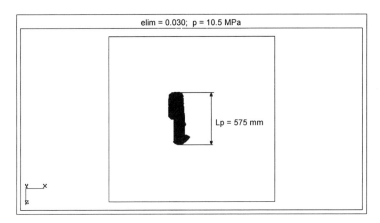

Figure 5: Model defect plastic area corresponding to $p = 10.5$ MPa and $\varepsilon_{LIM} = 0.03$.

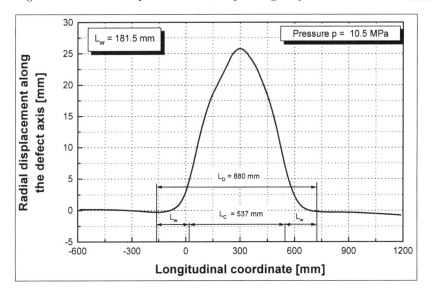

Figure 6: Graphical assessment of the defect core at $p = 10.5$ MPa.

The dependence of the defect relative plastic area length on pressure, i.e. $\{\Lambda, p\}$, is plotted in Figure 7 for different values of ε_{LIM}. Illustrated in Figure 7 are also the limit values Λ_{LIM} based on the authors' research results obtained as far for DN800 pipes, made of X60 steel (see the right upper corner). It can be seen that none of the curves (computed up to the pressure $p = 10.5$ MPa) intersects its limit value. This result could be expected with regard to the higher value of the yield stress of the St 52.3 steel comparing with X60. Nevertheless, the authors' results (DN800, X60) were further utilized to predict an approximate interval into which the limit pressure value of the DN900 pipe (made of St 52.3 steel) could be laid. Used for the prediction were the test results of the DN800 pipes, having machined artificial defects of the 360 mm in length, (Valenta, Michalec, Sochor, and Španiel 1992). Chosen from them were (Figure 8):

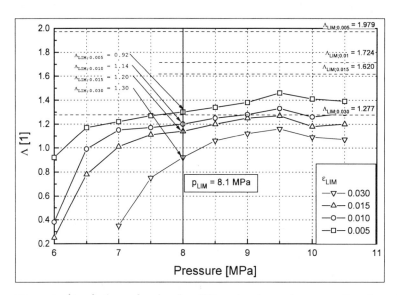

Figure 7: $\{\Lambda, p\}$ dependencies for different values of ε_{LIM}. Based on the model body actual burst pressure $p_{LIM} = 8.1$ MPa, indicated are corresponding values of Λ_{LIM} for the DN900-St52.3 model body. Λ_{LIM} for the DN800-X60 pipes are plotted in the right upper corner.

Defect	min. thickness [mm]	limit pressure [MPa]
I	3.7	9.5
II	4.5	11.72

These both defects had already been computed (using the authors' original program FEM211). Used from these computed results were values of the circumferential (σ_c) and equivalent (σ_e) peak stresses in the defects in dependence on several chosen pressure values and the burst pressures. These values were compared with a fictive theoretical stress state (σ_f) resulting from considering the DN800 pipe having its wall reduced to the defect minimum thickness. In this way, determined were coefficients $k^c_{800} = \sigma_c/\sigma_f$, $k^e_{800} = \sigma_e/\sigma_f$, respectively. For the model defect ($t_{min} = 4$ mm) machined on the DN900 pipe, similar coefficients $k^{c,e}_{900}$ were assessed (applying both the FEM elastic-plastic and analytical computations).

Plotted in Figure 8 is a constructed dependence of the k magnitude on the pressure. As the dependence appears to be approximately linear, this was substituted by linear regression. Based on the limit pressures of the comparative (DN800) areal defects (determined by past experiments), assessed were critical values of the coefficients $k^{c,e}_{800}$, being $k^c_{crit} = 0.78$, $k^e_{crit} = 0.84$, respectively.

Presuming a similarity in the limit states of the compared areal defects, similar values of their critical coefficients k_{crit} could be expected. Thus, the carried-out extrapolation of the coefficient k_{900} dependence on the pressure as far as into a limit region (Figure 10) enabled predicting the limit pressure for the model defect. The two regression straight lines for the coefficient k_{900}, plotted in the Figure 10, resulted from applying:

a) the preliminary computation as only as to $p = 4$ MPa (the CM4 line)

b) the further carried-out computation up to $p = 8$ MPa (the FM4 line)

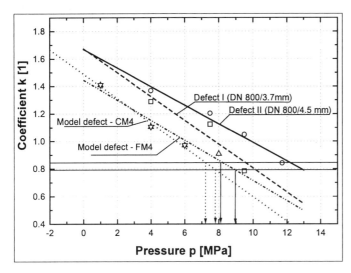

Figure 8: Regression and extrapolation of the FEM data obtained for the model defect (DN900-St52.3) and two preceding artificial defects (DN800-X60).

Examining the intersections of the CM4 line with the horizontal lines corresponding to k_{crit}, a preliminary limit pressure interval: $p_{LIM} = \langle 7.2; 7.8 \rangle$ MPa, was estimated, whereas from the FM4 line there resulted a new interval: $p_{LIM} = \langle 8.1; 9.0 \rangle$ MPa. The experiment on the model body determined the burst (i.e. limit) pressure being $p_{LIM} = 8.1$ MPa, and thus confirmed that this methodology may deliver a good quality prediction. Furthermore, the computed dependencies $\{\Lambda, p\}$, plotted above in Figure 5, in combination with the experimentally obtained $p_{LIM} = 8.1$ MPa, delivered very valuable parameters for the limit analysis of the DN900-St52.3 pipes, namely the limit relative plastic area lengths of defect Λ_{LIM}. In a further assessment procedure, the parameters Λ_{LIM} served to a satisfactory defect minimum thickness t_{min} estimation for the broken-down pipeline.

As the DN900-St52.3 pipeline ruptured at the pressure of 5.6 MPa, and the model body (though having an adequate size and shape of the model defect) burst at a markedly higher pressure ($p_{LIM} = 8.1$ MPa), it was evident that the model defect minimum thickness $t_{min} = 4$ mm was overestimated. This such that the model defect minimum thickness was reduced to be $t_{min} = 3$ mm and was denoted as FM3. The FM3 model was derived from the FM4 model by reducing the remaining defect thickness from 4 mm to 3 mm in all the corresponding identification points. Resulting was a mesh having 350 isoparametric elements with 1868 midside nodes representing 5604 degrees of freedom. Since in this case both the defect identification and FEM meshing errors did not in their geometry exceed the value of 1 mm, assessed was the lower boundary of the limit pressure estimation. The new limit pressure prediction, based on the FM3 model, utilized the experimentally assessed value $\Lambda_{LIM} = 1.20$ (being determined for $\varepsilon_{LIM} = 0.015$ and $p_{LIM} = 8$ MPa, Figure 5).

According to the limit pressure assessment explained above (using the FM4 model) computed were (for the FM3 model) analogical quantities (their dependencies on the pressure being plotted in series of similar figures) which, e.g., for $\varepsilon_{LIM} = 0.015$, $p_{LIM} = 8$ MPa, delivered: $L_P = 552.5$ mm; $L_D = 856$ mm; $L_W = 181.5$ mm;

57

from which

$$L_C = L_D - 2 \cdot L_W = 493 \text{ mm} \tag{4}$$

and the relative plastic area length of the defect was found to be

$$\Lambda = \frac{L_P}{L_C} = \frac{552.5}{493} = 1.12 \tag{5}$$

In such way the obtained dependence $\{\Lambda, p\}$, plotted in Figure 9, delivered the limit pressure estimation (for the DN900-St52.3 model body while utilizing $\Lambda_{LIM} = 1.20$) to be $p_{LIM} = 5.95$ MPa.

This result coincides nicely with the actual operational pressure (5.6 MPa) of the broken-down pipeline.

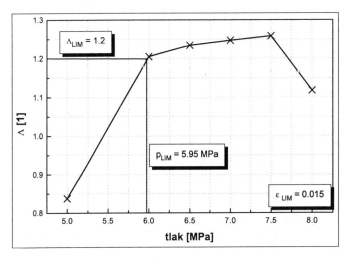

Figure 9: Graphical assessment of the limit pressure p_{LIM}.

4 ANSI/ASME STANDARD APPLICATION

The theoretical and experimental investigation was completed by the ANSI/ASME standard application. Based on numerous past experiences, a number of the by ANSI/ASME hitherto evaluated data (obtained by means of practically all its derived methodologies, e.g.: the basic and modified B31.G criteria, and the effective surface criterion) have been proved to be considerably conservative. Application of these methodologies for a different type of evaluation (e.g. for the limit pressure prediction) than that of the routine statements proving a potential satisfaction of the ANSI/ASME standard conditions, could seem not to have any other practical meaning. But just such theoretically predicted results (which have been verified by the experimentally obtained data) enable extending hitherto existing bases and thus gradually improving the precision of these standard procedures.

For the model defect evaluation, a pertinent basic profile of the model defect was to be assessed. For the purpose, a longitudinal section of the model defect was taken as the basic profile, whereof having a total length of 560 mm and maximum wall thickness reduction of 8.4 mm (i.e. minimum wall thickness $t_{min} = 4$ mm); see the last graph in the Figure 3 showing the defect longitudinal section in Y = 210 mm. The evaluation was carried out by

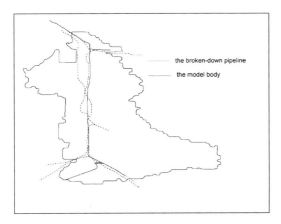

the broken-down pipeline
the model body

Figure 10: Comparison of the *broken-down pipeline/model defect* fracture lines.

the RSTRENG and CSTRENG211 programs, respectively. The CSTRENG211 program –
a CTU software product – enables the computational procedures to execute various set-ups
and create further variants. All applied standard variants predicted the model defect limit
pressure p_{LIM} being lower than the experimentally obtained burst pressure:

1. the basic B31.G ... $p_{LIM} = 4.5$ MPa

2. the modified criterion "85% defect profile area" ... $p_{LIM} = 7.21$ MPa

3. the effective area method ... $p_{LIM} = 6.1$ MPa

Surprisingly, the closest result was not delivered by the effective area method (outgoing
from a detailed topology of a defect and obtaining its computational depth and effective
length on the basis of the algorithm that divides the defect profile into a number of
strips, which being summarized step-by-step at different levels of the defect profile depth).
Rather, however, by a modified method considering the defect computational area as the
85% part of the rectangular area given by the multiplication of the defect length of 560 mm
and depth of 8.4 mm (Figure 3). The cause here consisted in the fact that the model
defect profile was not as articulated as it usually is with a typical pitting corrosion, but
that the defect profile virtually formed a rectangle (see Figure 3). It is obvious, according
to all the three applied criteria, that the allowable working pressure was assessed to be
substantionally lower and that this model defect was inapplicable for operation.

5 CONCLUSION

The breakdown cause of a DN900-St52.3 transmission gas pipeline was to be assessed. The
problem undertaken in this paper is based on the methodology of the CTU research team,
which being especially developed for the limit analysis of gas pipelines having corrosion
defects. Reconstructing and examining the pipeline locality where the breakdown took
place, found was a large areal corrosion defect that had initialized the accident. From a
number of distorted pipeline parts, reconstructed were the size and shape of the responsible
actual defect (AD) and, subsequently, according to the found AD dimensions designed and
manufactured was a model defect (MD). Because of the AD having undergone an extent
plastic flow in the course of the rupture process (resulting in a progressive contraction of
the AD remaining wall thickness t), the main problem consisted in a realistic assessment
of the original t_{min} (i.e. that before the rupture started).

Finally, based on a number of preliminary computations, the model defect minimum thickness was assessed to be $t_{min} = 4$ mm , with the model defect being machined on the model body (made of a DN900-St52.3 pipe taken from the accident vicinity). A nice similarity of the MD to the AD – and not only in their geometry but as well as in their operational behaviour – can be manifested by a close concurrence of their fracture lines (Figure 10) and a coincidence of the fracture onset locality of the both defects.

These facts can also be demonstrated when comparing the photo of the composed AD fragments (Figure 1(b)) with the photo of the MD taken after the burst test (Figure 5– where the fracture onset is situated near the KI4 strain gauge cross placement). The MD theoretical analyses, by application of the FEM211 software, were verified by experiments consisting in the model body hydrostatic burst test and connected with the strain gauge measurement of the MD deformations. In Figure 12, delivered is an illustration of the dependence of both the circumferential and longitudinal strains upon the model body pressurizing. The plots show how closely the theoretical deformation assessments applications – represented by the FM4 (the FEM fine mesh model with $t_{min} = 4$ mm) and CM4 (the coarse mesh model) – correspond with the nature of things. The graphs were based on the KI4 strain gauge cross signals, i.e. in the locality where the MD fracture was initialized.

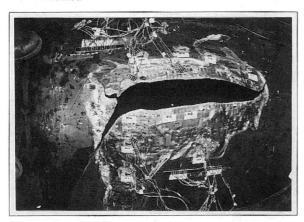

Figure 11: Model defect fracture after the burst test execution.

From the ANSI/ASME-B31.G evaluation modifications, the criterion "85% defect profile area" (delivering $p_{LIM} = 7.21$ MPa) reached very closely the limit pressure values obtained by both the FEM211 theoretical and experimental examinations.

Not only did the model body burst test ($p_{LIM} = 8.1$ MPa) verify the theoretical prediction of the model body limit pressure (9.2 MPa) as being quite satisfactory (though it was originally based on the DN800-X60 pipe research) but, at the same time, it delivered a limit value $\Lambda_{LIM} = 1.20$ being necessary for a qualified limit pressure prediction for the DN900-St52.3 pipes. Utilizing this $\Lambda_{LIM} = 1.20$ (meaning the relative plastic area limit length of the defect) and using FM3 computational model, assessed was the theoretical limit pressure ($p_{LIM} = 5.95$ MPa) of the model defect where having had $t_{min} = 3$ mm. This limit pressure estimation corresponded nicely with the operational pressure of the broken-down pipeline. The model defect susceptibility to the alternation of its minimum remaining thickness completes the view on the degree of its unreliability, whereby entailing that the operational safety of the pipeline having this kind of defect could be endangered with even a slight reduction of its wall thickness.

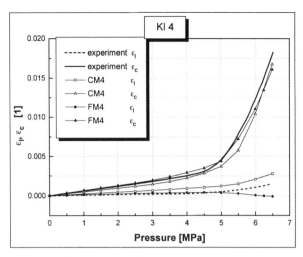

Figure 12: Comparison of the theoretically and experimentally obtained strain-pressure dependencies.

Having gathered all the above mentioned theoretical and experimental results completed with the pipeline material metallographic examination, the following concluding decision about the pipeline breakdown cause was stated:

1. the model defect shape and dimensions were proved to have been determined very realistic

2. comparing the difference between the broken-down pipeline pressure and the model defect pressure (while considering the FM3 model) of $\Delta p = 8.1 - 5.6 = 2.5$ MPa, the corrosion defect dimensions could not have been in its limit state, i.e. the DN900-St52.3 defect minimum remaining thickness could not have reduced down to as thin as $t_{min} = 3$ mm , from which there follows that our model defect thickness $t_{min} = 4$ mm corresponded nicely with reality

3. obviously, other influences played a role in the breakdown onset, e.g.

   (a) the pipeline bedding on an unyielding rock subsoil in the defect locality which could cause some insulation damage and, subsequently, the corrosion process onset

   (b) another possible influence for the breakdown onset could be non-homogeneity of material at the site of the defect line.

REFERENCES

Valenta, F., J. Michalec, M. Sochor, and M. Španiel (1990–1992). The theoretical and experimental stress analysis in surface defects of transit pipelines. CTU research works, CTU Prague, Fac. of Mech. Eng., Dept. of Elasticity and Strength of Materials.

Valenta, F., M. Sochor, M. Španiel, and J. Michalec (1994). Remaining load carrying capacity of gas pipelines damaged by surface corrosion. *International Journal of Pressure Vessels and Piping 59*(1-3), 217–226. ISSN 0308-0161.

Valenta, F., M. Sochor, M. Španiel, J. Michalec, M. Růžička, and V. Halamka (1996). Theoretical and experimental evaluation of the limit state of transit gas pipelines having corrosion defects. *International Journal of Pressure Vessels and Piping 66*, 187–198. ISSN 0308-0161.

Technical basis and software development for flaw assessments in NPP pipeline welds

A. I. Arzhaev, S. E. Bougaenko & I. N. Denisov
ECS MAE RDIPE, Moscow, Russia

V. V. Aladinsky & V. O. Makhanev
SRC NPO TSNIITMASH, Moscow, Russia

ABSTRACT: The basic parts of long-time strength maintenance are under consideration as applied to primary circuit piping welds. The use of technical, methodical and software developments providing fitness-for-purpose assessments is discussed.

1 INTRODUCTION

In recent years the concept of long-time strength maintenance of NPP components has been formed (Arzhaev et al. 1994). In accordance with this concept defect presence as well as its controlled growth are treated as the usual run of things during long-time service. Defect assessments are based on the fitness-for-purpose principle, determining that the defect would need to be repaired only if its presence is harmful to the integrity of structure. Practical realization of this principle is provided by development of new examination and life-time monitoring systems as well as advanced approaches to defect assessments. Some aspects of long-time strength maintenance concept are discussed below as applied to NPP primary circuit piping welds.

2 MECHANICAL PROPERTIES MONITORING

Mechanical properties monitoring is an integral part of strength maintenance. First of all, the properties related with the using methods of defect assessment (yield stress, fracture toughness, Jr-curves, etc.) are determined and predicted.

In accordance with approved technologies of assembly and repair welding the representative zones of weldment metal are picked out for which the set of necessary characteristics is determined. For example, as applied to primary circuit piping weldments it is possible to pick out up to seven representative zones as follows: pearlitic base metal, austenitic base cladding, pearlitic weld metal, fusion and heat affected zone of assembly weld, recovered austenitic cladding, deposited metal of repair weld, fusion and heat affected zone of repair weld.

It should be mentioned that the NPPs differ in primary circuit workpiece manufacturers and welding technologies. Therefore, defect assessments are executed using properties data for the specific NPP unit.

In order to determine the actual values of properties testings are carried out using weldment metal cut out during planned precautionary repairs after 100000 hours of operation. At present the experimental data corresponding 100000 hours of operation are collected and systematized for NPPs of first generation (Karzov & Timofeev 1995).

In order to confirm and obtain more specific information in application to the specific weldment containing defect non-destructive technologies of express-analysis are used. Micro-probes technology is used for microstructural analysis of metal embrittlement as well as for estimations of transition temperature and fracture toughness values. Micro-hardness probe technology is used to evaluate yield and ultimate stress levels.

3 ULTRASONIC EXAMINATION, DEFECT VISUALIZATION

Methods of ultrasonic examination are the main source of information on defect configuration and its growth. Last years modern ultrasonic holographic system «Augur 4.2» is used for expert inspection of NPP equipment (Arjaev et al. 1997).

Information supplied by «Augur 4.2» on precision defect dimensions corrects manual ultrasonic inspection data and has decisive influence on fracture assessments of detected defects. Results of «Augur 4.2» ultrasonic examination of main circuit pipeline weldments carried out at Russian NPPs and corresponding defect assessments show the following:

1) «Augur 4.2» examination procedure as well as the treatment of examination data should be conducted taking into account the recommendations of defect assessment method, which define guaranteed parameters of «safe» and «need to be repaired» defects;

2) in order to take advantage of «Augur 4.2» potentialities in full measure advanced methods of defect assessment should be applied using computer modelling and software developments.

4 ASSESSMENT METHOD ADAPTATION TO SPECIFIC STRUCTURE

At present, the M-02-91 method (RDIPE & VNIIAES 1991) is used in Russia for integrity assessments of structures with crack-like defects. This method is based on the interpolation approach which defines for defect under consideration the relationship between critical value of criterion characteristic and temperature. Under temperatures of brittle fracture the method uses the values of stress intensity factor and fracture toughness. Under ductile fracture conditions hardening effect is taken into consideration using values of limit plasticity and ultimate strength of metal.

Adaptation of M-02-91 method to main circuit pipeline weldments has been developed (Aladinsky et al. 1996). Defect location at the representative zone of weldment is analysed. Mechanical properties and metal characteristics for this zone are treated by the way providing conservative determination of assessment parameters. The results of finite element method modelling are used as applied to weldments with specific geometry and typical defect configurations. Interaction of defects located in parallel planes is analysed under the conditions of brittle and ductile fracture behaviour. Strain-stress fields near long-length defect with local increasing of its height are modelled to estimate this effect and determine scale factors. Residual stress distributions obtained by finite element method for assembly and repair weldments are used for precise analysis and margin assessments.

However, the regulations of M-02-91 method restrict potentialities provided by «Augur» system (defect configuration) and finite element analysis (stress-strain state of weldment). The advantages of adaptation and precise data are eliminated by high values of safety factors prescribed in M-02-91.

5 THERMOMECHANICAL HISTORY OF WELDMENT METAL

As applied to weldment metal its thermomechanical history is defined by non-isothermal low-cycle loading in accordance with bead deposition. Finite element modelling shows that weld root metal during multipass assembly welding is loaded by 15-20 cycles of plastic deformation with strain accumulation. Such pass-by-pass loading leads to the damage fast increasing during first passes and following damage growth in consequence of weld root strain concentration, weldment materials heterogeneity and strain ageing (Daunis et al. 1994).

Indirect partial incorporation of thermomechanical history by the use of residual stress distributions is insufficient for adequate in-service defect assessments. Two technological variants of welding were used for main circuit pipeline assembling. Finite element modelling showed that the technological variants resulted in the similar residual stress fields, but the weld metal (first of all, near the weld root zone) had different thermomechanical history and inherited damage (Aladinsky & Makhanev 1996). These differences had become apparent by crack-like defects starting from the weld root and detected during in-service examination of the circuit.

It can be seen that different NPP units operate under similar loading conditions. Therefore, defect acceptability is defined by defect configuration and inherited fracture resistance characteristics of metal with corresponding history of assembly and repair welding, heat treatment, operation. At present, accuracy of defect configuration measurements is provided by system of defect visualization («Augur 4.2»), but the approved method of defect assessment (M-02-91) had been developed and validated for estimations on the basis of current strain-stress state.

6 MODERN COMPUTER-BASED METHODS OF DEFECT ASSESSMENT

Strength maintenance, besides safe operation providing, directs researcher's attention to the following important problems:

1) determination of real safety factors conditioned by fracture resistance characteristics;

2) development of parametrical approach to the setting of in-service examination zones on the basis of defect ranging in accordance with the real safety factors.

Solutions of these problems may be founded on the way of new methodical developments using methods of computer modelling and analysis, but not on the way of tuning of present-day defect assessment procedures.

One of possible directions of the developments is common approach to local fracture description using local strains (local displacements) concept (Aladinsky & Makhanev 1995). The local strains (local displacements) concept is based on the consideration of small local volume with finite representative radius. Stress-strain fields inside the volume and associated fracture processes are completely defined by the local displacements: displacements of material points located on the surface of the representative volume. These displacements are determined by finite element modelling taking into consideration nonlinear behaviour of metal and thermomechanical history of loading. Tensor of local strains associated with the representative volume is introduced using integral functions of local displacements. Damage accumulation models are developed and fracture criteria are formulated in terms of local strains.

Local strains concept applications are inseparably linked with computer realization of the concept on the base of finite element method package: concept methodology and procedures of strength and fracture analysis are built in the package post-processor.

On the one hand, this approach provides continuous all-stages description of thermomechanical history and damage accumulation of weldment metal by finite element modelling and analysis. On the other hand, representative volume and local strains tensor define the procedure of transfer from specimen testing data to characteristics of welded structure containing defect.

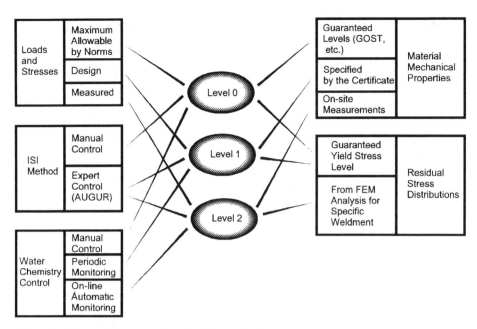

Figure 1. Scheme of conservatism levels introduction.

Realization of local strains concept is applied to main circuit pipeline weldments for fracture resistance analysis of typical heterogeneous problems: crack-like defects started from pearlitic weld metal and rested against austenitic cladding, or detected near austenitic repair weld boundary with underlaying pearlitic metal.

7 SOFTWARE DEVELOPMENT AND APPLICATION

Methods of computer modelling and analysis as well as software development are considered as important element of strength maintenance. This paper outlines in brief software for defect assessments in main circuit piping welds, developed as a integral part of strength monitoring system (Glazov et al. 1996) and realizing M-02-91 defect assessment procedure.

Set of crack-like defects under consideration includes internal elliptical, surface semi-elliptical and through thickness cracks. Long-length internal (under cladding) and surface crack zones are analysed by special way taking into consideration local increasing of defect height.

Assessment procedures, developed for defect type, include interpretation of ultrasonic examination data, characterization and re-characterization of defect, fatigue growth estimations, potential failure modes modelling, sensitivity analysis.

Information on operation loading, stress-strain state, material properties, detected defects is stored and treated by the databases of strength monitoring system. Data, obtained by different ways, differs in conformity with the real conditions. In order to provide coordinated assessments and ranging of defects with different data sets special approach is used. First of all, several relative levels of conservatism are introduced and input data sets corresponding these levels are defined (Figure 1). Then, based on experience of service maintenance, results of previous assessments and expert analysis, safety factors values are defined and brought into correlation with the levels of conservatism. By this way response functions are constructed to describe ordered behaviour of safety factors related with the input data margins relaxation.

66

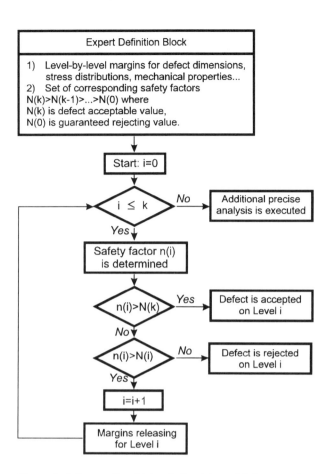

Expert Definition Block
1) Level-by-level margins for defect dimensions, stress distributions, mechanical properties...   2) Set of corresponding safety factors   N(k)>N(k-1)>...>N(0) where   N(k) is defect acceptable value,   N(0) is guaranteed rejecting value.

Start: i=0

i ≤ k — No → Additional precise analysis is executed

Yes

Safety factor n(i) is determined

n(i)>N(k) — Yes → Defect is accepted on Level i

No

n(i)>N(i) — No → Defect is rejected on Level i

Yes

i=i+1

Margins releasing for Level i

Figure 2. Scheme of **multi-level algorithm** of defect ranging.

Defect assessment and ranging are executed in accordance with multi-level algorithm by the way of step-by-step reducing of analysis conservatism under the condition of ordered safety factors increasing (Figure 2). Three safety factors with three response functions, corresponding normal operation, hydraulic testing and emergency conditions, are determined, and as a rule three levels of conservatism are used for practical evaluations.

At the stage of defect assessment any separate weldment is considered as independent object for analysis and maintenance. Necessary information for damaged weldment is extracted from databases and collected in the form of quasi-file with specific format by special program routine. Access to the quasi-file data is executed by the analogous program routine. Inside the defect assessment computer program the data are treated as a tables. By this way global databases and defect assessment software independence as well as network approach to the analysis are provided. Defect assessments are carried out for all weldments containing defects during limited period of time corresponding in-service inspection of NPP. In order to confirm lack of discrepancies, the last stage of assessment procedure consists in comparative analysis for all weldments and defects (for all quasi-files).

8 CONCLUSIONS

The fitness-for-purpose principle for defect assessment as applied to NPP pipeline welds is supported as follows:

1) methods of monitoring and express-analysis provide up-to-date information on mechanical properties and fracture characteristics;

2) ultrasonic system of defect visualization is used to get precise data on defect location and configuration;

3) computer modelling and analysis are used for adequate strain-stress estimations, as a base for advanced methods of defect assessment;

4) defect assessments are executed based on special methodical adaptations to real objects, software realizing assessment procedures is developed;

5) strength maintenance software provides well-timed and well-founded estimations, minimizes subjective factor influence, prevents loss or corruption of input-output data.

REFERENCES

Aladinsky, V.V. & V.O. Makhanev 1995. Computer system for strength analysis and residual service life evaluation of pipeline elements with damage and defects. *2nd Int. Cong. Protection-95, Moscow, 20-24 November 1995:* Abstracts, 51.

Aladinsky, V.V. & V.O. Makhanev 1996. D-800 weldments: residual stress modelling and its application to fracture analysis. *Proc. 4th Int. Conf. on Material Science Problems in NPP Equipment Production and Operation, St. Petersburg:* Vol. 1., 50-57.

Aladinsky, V.V., V.O. Makhanev, L.B. Babkin, A.I. Arzhaev & N.I. Karpunin 1996. Methodical aspects of acceptable defect assessments in D-800 pipeline weldments. *Workshop on defect assessments during NPP operation, 11-14 March 1996, St. Petersburg.*

Arzhaev, A.I., S.E. Bougaenko, I.N. Denisov & V.A. Kiselyov 1994. RBMK components strength-state monitoring system as an effective tool for NDE results assessment. *Joint IAEA, OECD & CEC Specialists' Meeting on Non-Destructive Examination Practice and Results, Petten, 8-10 March 1994.*

Arzhaev, A.I., V.A. Kiselyov, V.G. Badalyan, A.Kh. Vopilkin, V.N. Vanukov, V.V. Aladinsky & V.O. Makhanev 1997. Field application of «Augur» ultrasonic system during RBMK NPP unit ISI and its impact on pressure boundary integrity assessment. *Cape'97 — Proc. 4th Int.Colloq., Cape Town, 21-25 April 1997:* Rotterdam: Balkema.

Daunis, M., S. Dauniene & B.T. Timofeev 1994. Strain aging of low carbon steels and their weldments by low-cycle fatigue. *Proc. 3rd Int. Conf. on Material Science Problems in NPP Equipment Production and Operation, St. Petersburg:* Vol. 1, 101-113.

Glazov, O.N., A.I. Arzhaev & I.N. Denisov 1996. Prospect of RBMK-NPPs strength monitoring system on-line module development. *Int. J. Pres. Ves. & Piping.* 66: 381-386.

Karzov, G.P. & B.T. Timofeev 1995. Analysis of long-time effects of operation temperature and strain ageing on the mechanical properties of pressure vessels and pipings. *Appendix to RDIPE Report No 23.5671.* Moscow (in Russian).

RDIPE & VNIIAES 1991. Procedure for allowable flaw size evaluation in metal of equipment and piping during NPP operation (M-02-91). Moscow (in Russian).

Asset management

Risk based asset life management

J.R.Lilley & M.A.Davies
AEA Technology plc, Warrington, UK

ABSTRACT: Items of plant equipment are frequently inspected for statutory or economic reasons. Inspection regimes have been derived through ranking items on the basis of criticality which have been influenced and refined over the years through engineering knowledge and experience. This philosophy is sound in that it has a demonstrable track record of containing plant failures within the limits of practicality, but today we have access to a greater knowledge of the issues involved such that the safety and reliability of plant at factory, process or item level can be addressed on an individual basis, taking specific conditions into account to manage risk of failure throughout the lifetime of an asset within acceptable levels. The use of risk based methods in the preparation of fitness for purpose justifications for assets such as process plant has several potential benefits which include: the reduction of down-time by replacing intrusive by non-intrusive inspection and testing; the reduction of inspection and testing effort which has no significant influence on the risk; more extensive use of previous 'favourable' inspection and testing results and work already undertaken to support an overall facility Safety Case; and production of justifications in a format acceptable to regulators. A number of case studies will be discussed illustrating the AEA Technology approach to risk based asset life management.

1 INTRODUCTION

Industry wide life management practices naturally embody degrees of conservatism to account for the extremes of variation to be encountered in the contributors to plant failure. The selection of items for inspection also tends to be generic and little credit is given for the type of inspection carried out. As we have increasing access to plant and process performance information, inspection technology and monitoring tools have made significant advances, and as our understanding of flaw behaviour and materials performance is greatly enhanced, it is valid to challenge the conservatism in the conventional approach to plant life management. The risk-based approach to asset life management enable these contributors to risk to be identified and managed through the adoption of a three phase approach involving:

a) Risk-Based Inspection Planning

b) Inspection

c) Assessment

The objective is to identify the potential causes of failure and to produce a management strategy to target the identified failure mechanisms at appropriate points in time using suitably reliable inspection techniques to detect and size anomalies to contain the risk of failure within tolerable limits.

The Risk-Based inspection planning phase singles out the contributors to risk and produces an inspection plan defining interval, scope and type of inspection required using risk of failure to define when, where and how to inspect. Risk may be ranked qualitatively or quantitatively as conditions dictate. Inspection technology has developed to the point where a wide variation in inspection techniques are now available which give a greater choice of cost balanced against inspection reliability. Assessment techniques have also greatly advanced with computer models to predict material behaviour supported by proven probabilistic and statistical methodologies.

The conventional approach to plant inspection and life assessment remains to be the most viable method in most instances, but there are situations where a more detailed approach is justifiable. The approach can be applied at a high level to screen out the instances where it will not provide benefit and to focus on those where it is mostly applicable. Implementation of the approach requires a great deal of expertise and technology support which has cost implications, but when balanced against the significant benefits of improvements in the overall risk profile, plant availability and reliability these costs frequently prove to be justifiable.

2 RISK-BASED INSPECTION PLANNING

Risk has been defined as "the chance of something adverse happening" (HMSO 1989). It can be seen that "chance" refers to a frequency or probability and "something" refers to a consequence which, in general, could be either safety, economics or environmentally related. Thus we can write:

risk = frequency * consequences

The use of risk based methods requires an acceptable risk to defined. Since there is no absolute standard of acceptable risk this becomes subjective. The risk level that is acceptable is usually arrived at by interaction between a plant operator and the regulator. If the consequences of plant 'failure' are known and the acceptable risk has been agreed then by manipulation of the definition an acceptable failure frequency can be determined and used as a performance criteria.

The main objective of risk based methods from a safety point of view is to demonstrate that the risk to operators or the General Public from operation of the plant below this agreed acceptable level. The conventional approach to plant inspection and life assessment is via established codes and standards which can contain conservatisms because they apply to a wide range of industries. Thus if a critical component of the plant can be shown to satisfy the code then it is assumed that the associated level of risk will be acceptable and this is used to justify continued operation of the plant. This is not always the case and codes and standards have actually been shown to be non-conservative or recommended inspection methods have been shown to be inappropriate. Explicit assessment of the risk associated with plant operation should remove some of the uncertainties assocaited with prescribed rules and regulations. The implementation of risk based methods to indicate where, when and how to inspect and maintain could significantly extend the useful life of the plant or asset.

Many countries are responding to the above arguments and the regulations governing assets such as petrochemical plant are undergoing revision. The trend is away from regulations couched in prescriptive terms to a more goal oriented risk based approach. In the United kingdom, for example, both The Pressure Systems and Transportable Gas Containers Regulations 1989 (HSE Books CD92 1995) and The Pipelines Safety Regulations 1996 (SI 1996/825) (HSE Books DDE2 1995) embrace the goal setting approach. A number of benefits can be obtained from adopting the risk based methods of the type described in this paper. Some of these benefits are outlined below:

Reduction of Downtime or Lost Production - Prescriptive regulations often require, for example, periodic in-service intrusive testing involving vessel openings. Goal setting risk-based regulations allow justifications to be made for replacing such intrusive testing by appropriate and well-targeted non-intrusive testing. Such justifications can be made either on the basis of no increase in risk when the non-intrusive testing option is followed or on the basis that although the risk associated with the non-intrusive testing option may be greater it is still acceptable. The savings associated with replacing intrusive by non-intrusive testing are often substantial.

Reduced Inspection and Testing of Low Risk Items - It is clear that reduced levels of inspection and testing can be justified for low risk items. Direct savings in the costs of such inspections can be substantial.

More extensive Use of Existing Information - Prescriptive regulations often simply require periodic inspection and testing with no benefit being allowed for 'favourable' results. The risk-based methods described in this paper allow fuller use of existing inspection and testing data by constantly updating plant risk profiles when new information is obtained. For example, an initial risk assessment might identify a particular vessel as high risk, with corrosion being the dominant damage mechanism. Regular corrosion mapping might be specified for the vessel as a result of this assessment. If such regular monitoring in fact indicated a lower than anticipated corrosion rate this information would be used to extend the originally specified inspection interval on the basis that the risk was acceptable with the increased interval. Risk-based methods are also able to use many analyses undertaken as part of an overall facility Safety Case (e.g. fire and explosion assessments). This gives added value to such Safety Case work.

Regulatory Acceptance

It is clear that justifications based on an explicit consideration of risk will be naturally in a format which facilitates acceptance within a goal setting risk-based regulatory regime.

Risk Based Inspection Methodology

A typical high level safety justification strategy developed by AEA is shown in Figure 1. The first stage consists of hazard identification and qualitative risk ranking

(QRR). This was implemented by use of a structural HAZard and Operability study (HAZOP). The HAZOP study technique utilises a systematic and thorough approach which takes into account a wide range of relevant knowledge and experience of the plant. This stage identifies all hazards which might affect the structural integrity of the vessel and pipework. Each hazard is then assigned a qualitative likelihood and severity category in terms of personnel or General Public safety on a judgmental basis using the knowledge and experience of the HAZOP team.

The plant risk profile can then be formed into a risk matrix, shown in Figure 2, to enable the high risk hazards requiring more detailed assessment to be easily identified. This approach ensures that an inappropriately detailed level of analysis is not performed on items posing only a low risk.

The areas of plant considered to be non-negligible contributors to total risk identified were then subjected to appropriate NIT, such as Time of Flight Diffraction (TOFD) and Seescan also specified by the above approach. These inspection techniques will be discussed in more detail in the next section.

The non-negligible contributors to total risk along with the results from appropriate inspection are then addressed in detail in the Safety Justification. The assessment carried out in this phase of the risk based methodology is determined by the nature of the Safety Justification sought. This may include thermal hydraulic calculations, fracture mechanics assessment and analysis of pitting or uniform corrosion.

3 INSPECTION

Developments in inspection technology are continually extending and improving the range of tools available to the inspection engineer. Also, studies and evaluations of traditional Non-Destructive Testing methods and techniques improve our understanding of the reliability and confidence levels which can be applied. Once an

Figure 1 High Level Safety Justification Strategy

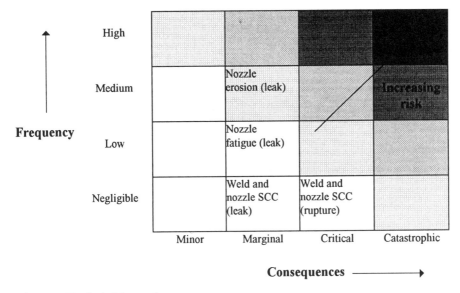

Figure 2 Typical risk matrix

inspection specification has been decided on, i.e. a definition of where, when and what to look for, we are able to design an inspection approach which will provide the required level of confidence.   There are large variations in the effectiveness of inspection techniques. A recent study by the Dutch Welding Institute (NIL) of Non-Destructive testing techniques for welds in ferritic steels in the thickness range 6mm to 15mm showed variations in the effectiveness of individual techniques and highlighted serious shortcomings in techniques which have over decades been assumed to be reliable (Stelwagen 1995).

Weld assessment - the NIL study

The purpose of the study was to evaluate the performance of a range of Non-Destructive Testing techniques in blind trials on 21 welded low carbon ferritic steel samples between 6mm and 15mm wall thickness containing 244 artificial, yet realistic flaws such as lack of penetration and fusion, slag and gas inclusions and cracks. The flaws were all longer than 10mm and were mostly between 1 and 4mm in height. The results of the radiographic techniques follow:

Table 1.   Flaw detection rates of radiographic techniques for various thickness of plate.

Technique	Detection rate (%)			
	6+8mm	8+10m m	10+12mm	15m m
X-Radiography	69.0	62.5	66.2	67.0
Gamma-Radiography	63.0	53.1	53.6	70.5
Bevel Radiography	93.5	94.0	96.0	95.0

 Source

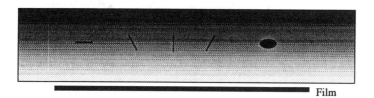
Film

Figure 3. Radiographic inspection, which is sensitive to variations in density of the item under examination, and commonly believed to be an effective technique is very insensitive to oblique planar flaws such as lack of fusion and cracks

The effect of flaw orientation with respect to beam direction is clearly evident. Bevel radiography is X-Radiography involving two exposures of each section of weld with the beam oriented along each fusion face. This is a laboratory technique and is not practical for field use due to intricate setting up/alignment and an effective doubling of testing duration. It should be borne in mind when evaluating the results that volumetric flaws such as slag and porosity are detectable by radiography irrespective of incident beam angle. The difference between the figures for X-Gamma-Radiography and Bevel Radiography are attributable to planar flaws. This is of major significance when it is considered that code acceptance criteria disallow lack of fusion and cracks!

Radiographic techniques record differences in the efficiency of transmission of ionising radiation passing through the item being tested. More energy is transmitted through less dense areas of welds such as localised gas or slag entrapment, but the transmission is relatively unaffected by thin, inappropriately oriented flaws such as cracks and fusion flaws. Radiography is therefore very good at detecting volumetric flaws and poor at detecting planar flaws (Figure 3). Acceptance criteria used in practice assign strict limits on the allowable sizes and distribution of volumetric flaws and generally disallow fusion flaws or cracks. Yet the process of excavation of volumetric flaws and subsequent repair welding probably causes more damage through weakening the metallurgical structure than if the flaw were left intact and the radiographic inspection technique often specified by the code is clearly unsuited to the task of planar flaw detection. The propensity of radiography to cause damaging repair of volumetric flaws and its inherent inability to detect most planar flaws begs the question of the relevance of the routinely applied code criteria in combination with radiography.

The study also showed that variations in performance are apparent if different procedures using the same technique are applied. For instance, the Time-of-Flight-Diffraction technique (TOFD) exhibited variations in detection rates between 45% and 97.7% (Table 2).

Table 2. Flaw detection rates for a range of Time-of-Flight-Diffraction techniques for various thickness of plate.

Technique	Detection rate (%)			
	6+8mm	8+10mm	10+12mm	15mm
TOFD (1)	80.0	79.0	75.0	95.5
TOFD (2)	65.0	76.0	77.6	97.7
TOFD (3)	45.0	54.0	63.0	72.7

A detailed description of the Time-of-Flight-Diffraction technique will not be entered into here as it is described elsewhere (Charlesworth & Temple 1993). However, in simplistic terms, where ultrasonic technology in the medical field images internal features of the human body, the time-of-flight-diffraction technique images the internal structure of metallic items such as welds. The ability of the technique to cover the weld, heat affected zone and adjacent parent material very quickly with a single ultrasonic pulse and to detect flaws of any orientation within this volume are major attributes which have significant advantages for the testing of welds. Appropriate instrumentation with low electronic noise characteristics and adequate imaging technology enables the image to display acoustic scatter from the material structure itself. This is therefore a highly sensitive technique and it enables weak signals generated at the tips of very tight fusion flaws and cracks to be detected and accurately located. By traversing a probe pair along a weld at speeds of 100mm/s an image of the internal structure is generated in real time. The ultrasonic signals are digitised and stored for subsequent retrieval and analysis if required.

The three time-of-flight-diffraction procedures evaluated in the study clearly exhibit variations in performance. The differences lie in the instrumentation itself, the procedures adopted and the calibre of personnel. Each of these performance criteria can however be quantified, specified and measured through validation, certification and the adoption of appropriate standards.

The study provided valuable source information on the overall capabilities of a wide range of commercially available Non-Destructive testing procedures from which we can derive information which can be fed directly into a risk based inspection programme. For instance, Table 3 shows a significant difference between the performance of automated and manual ultrasonic testing procedures. The two procedures were identical except the automated procedure used a mechanised scanner

Table 3. Flaw detection rates for a range of techniques for various thickness of plate.

Technique	Detection rate (%)			
	6+8mm	8+10mm	10+12mm	15mm
X-Radiography	69.0	62.5	66.2	67.0
Gamma-Radiography	63.0	53.1	53.6	70.5
TOFD (1)	80.0	79.0	75.0	95.5
Automated UT (1)	82.0	84.0	82.0	86.4
Manual UT	46.0	46.1	47.9	69.0

to manipulate the probes and the data was digitally captured, stored and displayed using imaging technology whereas the manual procedure relied on manual probe manipulation and on-line interpretation of analogue signals. The variations between the two procedures are due to human reliability factors and risk of plant failure can be improved through automation of manual procedures.

Through investigation of the performance of each procedure evaluated in the study, further improvements may be attainable if inspection reliability needs to be extended. For example, the time-of-flight-diffraction technique was found to be unsuited to the detection of excess penetration in single sided welds and was ineffective at the detection of flaws located away from the immediate vicinity of the weld, heat affected zone and adjacent parent material. In the event that these anomalies are a concern they can be addressed by internal visual inspection in the case of excess penetration and additional scanning in the case of flaws remote from the weld region. If these flaw types are accounted for in this manner, the detection rate for time-of-flight-diffraction increases to over 95% for all thicknesses. It should also be taken into account that the thickness range covered represents the lower end of the thickness range of the application of the time-of-flight-diffraction technique. Performance increases with thickness (Charlesworth & Temple 1993).

Corrosion assessment

Vessels and pipes are frequently assessed in practice for material loss through corrosion by internal visual inspection and/or point measurements by ultrasonic means. Both approaches have limitations in that they can (and frequently do) fail to detect localised areas of corrosion due to factors such as vessel internals being present which restrict access, corrosion product, debris and contaminants masking corroded areas, the sheer volume of extensively pitted areas making it difficult to identify the worst case locations, and in the case of point, or grid thickness readings the volume of effective coverage achievable in practice represents a tiny percentage of the total surface area. Furthermore, visual inspection is subjective in nature and provides no hard-copy evidence of material condition.

An alternative approach to corrosion assessment which is complimentary to the risk-based approach to asset life management is corrosion mapping. This is where an ultrasonic probe is driven over the area of interest from the outside surface of the component in a raster formation taking wall thickness measurements at discrete intervals, typically in steps of a few millimetres. These thickness measurements are colour-coded to represent the various thicknesses of material and are plotted on a position-related image. The data is digitally stored and can be retrieved for analysis/processing off line. See figure 4.

The interval between individual ultrasonic measurements dictates the resolution of the resultant image. For generalised erosion without any sudden changes in thickness, a coarse resolution may be adequate, but to identify and analyse specific forms of complex types of corrosion a fine resolution may be required (refer to (Stelwagen 1995) for further information on the selection of inspection resolution). High resolution ultrasonic mapping can enable corrosion type to be identified and in certain cases this information can be used to underpin a remedial strategy to inhibit the rate of material loss. The effectiveness of inhibition strategies can be monitored by repeating the inspection at intervals.

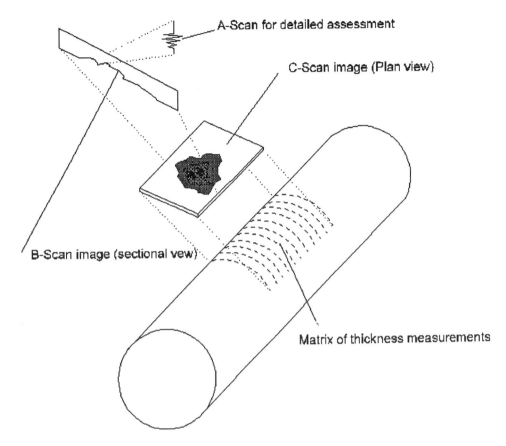

A-Scan for detailed assessment

C-Scan image (Plan view)

B-Scan image (sectional vew)

Matrix of thickness measurements

Figure 4. Corrosion mapping and the benefit of ultrasonic Waveform Capture.

Ultrasonic corrosion mapping provides detailed images of component internal condition. Individual scans, when assembled to form a composite image represent a sufficiently large area to portray the influence of product flow, process conditions and the interaction of one area of corrosion with another.

Types of corrosion identifiable from composite maps include:
- under-scale corrosion
- erosion along the bottom of pipelines caused by sand pushed along with the product flow
- erosion caused by turbulence at areas of flow diversion such as risers, manifolds and tees
- corrosion at low points caused by water drop-out
- corrosion at areas of water entrapment following hydro-testing
- CO_2 corrosion pitting
- Hydrogen Induced Cracks

A further benefit of corrosion mapping is the ability to store digitised ultrasonic waveforms. This enables a greater flexibility in Off-Line imaging than where only thickness measurements are concerned. The procedure followed is to create a composite image of the area of concern, then select worst case or representative sections for detailed analysis. Individual B-scans are then analysed to study the influence of the reflector's morphology on the ultrasonic beam (see fig 4).

This can give a great deal of information on the mechanisms taking place within the material itself and at the inner surface. This process is useful for example, in discrimination between inclusions, hydrogen induced cracking and corrosion pits and for assessing whether hydrogen induced cracks are 'rough-surfaced' or inter-linked by step-wise cracking.

4 CASE STUDIES

Hydrotest Deferral

AEA Technology has prepared a Safety Justification for extending the hydrotest interval of a Waste Heat Boiler (WHB). The WHB forms part of a large process plant. The WHB cools gas at a temperature of 960°C from a secondary reformer unit to a temperature of about 520°C. To comply with regulations the water side of the WHB was hydro-tested at intervals not exceeding 15 months at a pressure of 1.23 times the design pressure. Because of the locations of check valves, other vessels and pipework were pressurised during each WHB water side hydrotest. Figure 5 is a schematic with the WHB hydrotest envelope shaded. The equipment concerned had suffered extensive metal dusting damage, particularly on the inlet tubesheet of the WHB. During previous hydrotesting small tube leaks had also occurred in the steam superheater.

The client wished to apply for exemption from an upcoming WHB water side hydrotest. AEA Technology were asked to write a Safety Justification to support this application. The three types of argument used in the Safety Justification to do this were:
(a)   Hydrotesting is not beneficial, therefore its deferral does not increase the risk.
(b)   Hydrotesting is beneficial but this benefit can be replaced by other means: for example, additional targeted inspection.
(c)   Detailed analysis, rather than qualitative risk assessment, can be used to show that the risk associated with a particular situation is in fact negligible, even with hydrotest deferral.

In relation to (c), full use was made of work already undertaken on this equipment. This work included finite element analyses of the WHB tubesheet and thermal hydraulic calculations of the safety consequences of a 'worst case' failure in the WHB tubesheet. Before inclusion in the Safety Justification this work was subject to independent Safety Assessment. This was done to ensure that the conclusions of such work were adequately robust for inclusion in the Safety Justification. This approach gave very significant added value to work already undertaken on the WHBs.

The qualitative risk assessment process identified eight non-negligible contributors to the total risk. These non-negligible contributors included steam drum nozzle erosion, WHB tubes metal dusting, WHB tubesheet metal dusting and steam superheater header fatigue. As noted previously, all non-negligible contributors to the total risk were addressed in detail in the Safety Justification. These detailed assessments used a combination of AEA Technology's own analyses and, after Safety Assessment, analyses already completed on this equipment. An example of AEA

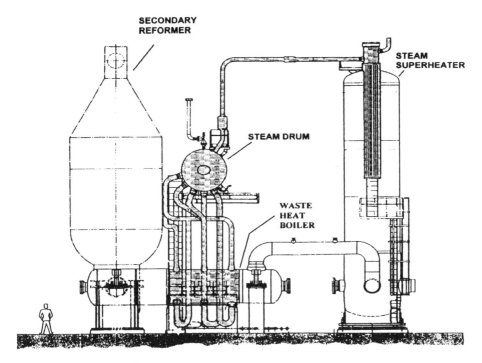

Figure 5 Hydrotest envelope

Technology's own analysis was thermodynamic modelling of the effects of coating the WHB tubesheet in an attempt to reduce metal dusting rates.

An exemption from the upcoming WHB water side hydrotest was granted.

Justification of a Gas Treatment Centre and Well Sites

AEA Technology prepared a Safety Justification for the continued operation of a gas treatment centre and well sites. The project was completed in three phases. Phase I involved preparation of a risk-based inspection plan for the vessels and pipework concerned. The inspection plan specified non-intrusive, rather than the traditionally applied intrusive, testing techniques. During Phase II AEA Sonomatic implemented the specified inspection plan. This involved ultrasonic Time of Flight Diffraction testing of selected vessel and pipework welds, and ultrasonic Seescan corrosion mapping of selected areas of pipework and vessels. Phase III involved preparation of the Safety Justification making full use of the Phase II inspection results. Phase III analysis involved a PD 6493: 1991 (Ref 7) fracture mechanics assessment of weld root erosion and an assessment of the significance of wall thinning produced by uniform corrosion. The analyses justified very long remanent lives for most of the plant. Additional monitoring or replacement was suggested for a few localised areas with comparatively short remanent lives. Because of the non-intrusive nature of the specified inspections, the plant down time for inspection was much reduced in comparison to that for an intrusive inspection.

AEA Technology have prepared a risk-based inspection plan for a naptha hydrotreater unit. A notable feature of this project was that Stress Corrosion Cracking (SCC) in stainless steel pipework (due to the leaching of chlorides from lagging) was identified as a damage mechanism with non-negligible risk. It had not been considered previously. This led to the client initiating a plant-wide programme to inspect for this type of damage. The programme detected this type of SCC. Early recognition of unanticipated forms of damage such as this is of great value to the safe and economic running of a plant.

5 CONCLUSIONS

Substantial benefits can be obtained by applying risk based methodologies in the production of fitness for purpose justifications for assets such as process plant. Risk based methods have a wide range of applications and are being used increasingly throughout industry.

If risk is to be managed throughout the life of plant, the inspection procedures adopted must be appropriate to degradation mechanism present. Failure to do so will increase the risk of failure. Tools are available to assist in the selection of appropriate inspection procedures such that inspection probability of detection (PoD) values can be linked to the probability of failure.

The ability afforded by inspection technology to provide information on flaw morphology, precise sizing information, improved probability of detection, hard copy evidence of coverage and to support and monitor remedial strategies all have the potential to improve the overall risk profile of plant.

A number of case studies have been discussed which illustrate how the potential benefits available through the use of risk-based methods have been realised in practice by a number of clients. Factors such as (i) the trend away from regulations couched in largely prescriptive terms to more goal setting risk-based approaches and (ii) the continuing pressures on plant managers to maximise the return from their assets means that risk-based methods of the type discussed in this paper are likely to become increasingly widely used. Such methods offer a rational framework for reducing costs without compromising safety standards.

REFERENCES

"The Pressure Systems and Transportable Gas Containers Regulations 1989", Her Majesty's Stationery Office, Statutory Instruments 1989 No. 2169, 1989.

"Proposed Pipelines Safety (PSR) Regulations 199-", HSE Books, Consultative Document CD92, 1995.

"Generic Terms and Concepts in the Assessment and Regulation of Industrial Risks", HSE Books, Discussion Document DDE2, 1995.

"NIL Project: Non Destructive Testing of Thin Plate" doc.no.:NDP 93-40 (Stelwagen. March 1995)

"Practical Applications of Time-of-Flight-Diffraction" (Charlesworth & Temple. 1993)

NDT - Value for Money. M Wall and F A Wedgwood. INSIGHT Vol 36 No 10 October 1994.

"Guidance on Methods for Assessing the Acceptability of Flaws in Fusion Welded Structures", BSi Standards, Published Document PD 6493, 1991.

Ageing of Materials and Methods for the Assessment of Lifetimes of Engineering Plant, Penny (ed.)
© *1997 Balkema, Rotterdam, ISBN 90 5410 874 6*

Probabilistic risk analysis of ageing components which fail on demand – A Bayesian model: Application to maintenance optimization of diesel engine linings

C.A.Clarotti
ENEA-INN-TEC, CR Casaccia, Rome, Italy

A.Lannoy
EdF-SDM, Chatou, Paris, France

H.Procaccia
EdF-REME, Chatou, Paris, France

ABSTRACT: This paper deals with deriving and discussing a probability model for optimizing the maintenance policy relative to stand by components, such as diesel generators and breakers, which fail on demand and age. These components are relevant to the safety of Nuclear Power Plants (NPP) and the main problem with them is the optimization of the maintenance and the test-schedule. Indeed the effect of any test is twofold, namely: the test reveals failures which otherwise would occur when the Power Plant would call for aid but it causes the aging process to step up. The aging process is suspected to affect certain safeguard equipments, but it does not exist a theoretical model of component aging in function of the demands. Sections 2, 2.1 and 2.2 of the paper are respectively devoted to the description of a Bayesian model of discrete aging, deriving the likelihood function of the model (data modelling) and to selecting "the best prior" (Analyst knowledge) that can be associated to the likelihood. The aspects of numerical integration relative to assessing the predictive reliability of the aging component are discussed in Section 2.3. Section 3 reports a RCM application of the model to components of Electricité de France (EDF)-Power-Plants.

1 INTRODUCTION

Reliability Centred Maintenance (RCM) is a rational approach allowing to identify the equipments and sub components of Power Plants that may turn out to be critical with respect to safety, to availability or to maintenance costs: only for these critical equipments will be established a policy of corrective or preventive maintenance.

The chosen maintenance policy strongly depends on the possibility of aging of the equipment under consideration, and of the consequence of one eventual failure, in term of risk or in term of cost.

Component aging with time is well modeled with a two parameters Weibull Probability Density Function (PDF). Unfortunately it does not exist a model of aging for components which fail rarely and which fail essentially on demand. This case corresponds to the aging process of diesel engine linings (cylinders).

The diesel generators of NPP are routinely tested (25 time per year), during short duration (1 hour) in average, because they are very seldom called into mission in case of real loss of power incident: loss of both external power supply sources (400 and 250 kV), with the conjunction of unsuccess of plant house loading (plant self supply from the main turbo-generator).

With these particular conditions, plant operators suspect that the aging of the diesel engine linings is essentially due to the number of starts up and not due to the test operating durations.

On the second hand, and fortunately, failures of diesel linings are rare events and it is difficult to model their probability of occurrence. Bayesian approach which allows to joint two sources of information: expert or analyst Prior knowledge with observed data (Likelihood), to obtain Posterior enriched probability, seems the best approach to use in case of rare events.

So in the following, we shall present a specific and original theoretical model making it possible to determine aging of equipments which fail essentially on demand, such as diesel engine linings, breakers, auxiliary pumps or generators....
One example relative to the determination of optimal preventive maintenance periodicity will be given after.

2 DERIVING THE MODEL OF AGING ON DEMAND

2.1 Modeling the failure observations : the Likelihood Function

2.1.1 Probability to observe the first failure at demand n

The model developed (Clarotti, Lannoy & Procaccia, 1996) stems from the following ideas.
1. Consider an urn with white and black balls and the gamble where the success consists in drawing a white ball.
2. If after any success the white ball is put again into the urn, the odds of the gamble remain constant.
3. If the total number of balls (white balls + black ones) goes to the infinity, the Bernoulli model is obtained.
4. The Bernoulli model is used for components which fail on demand and do not age (constant probability of failure on demand).
5. The model for components which age can then be obtained by letting the total number of balls go to the infinity in the case that after each success one puts into the urn the white ball which has been drawn together with v black balls.
   Consider the urn model in the case that the number of black balls is increased at any drawing of a white ball. Let N and k respectively be the total number of balls and the number of black balls at the beginning of the gamble. If the values of N and k are known, the conditional probability $\lambda(n|\theta, \delta)$ that a black ball is drawn for the first time at the n-th step given that no black ball has been drawn before, is clearly

$$\lambda(n|\theta, \delta) = \frac{k + (n-1)v}{N + (n-1)v} \qquad (1)$$

Posing : $\theta = \dfrac{k}{N}, \delta = \dfrac{v}{N}$,

we obtain : $\lambda(n|\theta, \delta) = \dfrac{\theta + (n-1)\delta}{1 + (n-1)\delta},$ \qquad (2)

The total number of balls being $N + (n-1)$ v, and the total number of black balls being $k + (n-1)$ v.
It is easily shown (Clarotti, Lannoy & Procaccia, 1996) that $\lambda(n|\theta, \delta)$ is an increasing function of n, whose value lies between 0 and 1 for any (θ, δ) such that $(\theta, \delta) \in [0,1] \times [0, +\infty)$. By comparison with the Weibull distribution the function $\lambda(n|\theta, \delta)$ can then be interpreted as the failure rate at time n (i.e. the conditional probability of failure given the survival till the discrete time n) of a component which ages and fails on demand. The parameter θ is "the unknown probability of failure on the first demand" and δ quantifies the importance of the aging phenomenon. Let us denote by N_f the random (discrete) equivalent time to failure, say of a diesel engine or of a breaker. In view of the above we can write the Probability of failure and the complementary Probability of success

$$Pr\{N_f = n \,|\, N_f > n-1, \theta, \delta\} = \frac{\theta + (n-1)\delta}{1 + (n-1)\delta} \qquad (3)$$

$$Pr\{N_f > n \mid N_f > n-1, \theta, \delta\} = 1 - \frac{\theta + (n-1)\delta}{1+(n-1)\delta} = \frac{(1-\theta)}{1+(n-1)\delta} \tag{4}$$

Conditional on the knowledge of the parameters θ and δ, the probability that the component will survive the first n demands can be calculated according to eq. (5)

$$Pr\{N_f > n \mid \theta, \delta\} = Pr\left\{\left(\bigcap_{j=1}^{n}(N_f > j)\right)\middle|\theta, \delta\right\} =$$

$$= Pr\{N_f > 1 \mid \theta, \delta\}\cdot\prod_{j=1}^{n-1}Pr\left\{N_f > j+1\middle|\left(\bigcap_{i=1}^{j}(N_f > i)\right), \theta, \delta\right\} = \tag{5}$$

$$= Pr\{N_f > 1 \mid \theta, \delta\}\cdot\prod_{j=1}^{n-1}Pr\{N_f > j+1 \mid N_f > j, \theta, \delta\}$$

Indeed:
1. the statement "N_f is larger than n" is equivalent to the statement "N_f is larger than any

   integer whose value is not larger than n"; so events $(Nd > n)$ and $\left[\bigcap_{j=1}^{n}(Nd > j)\right]$ are

   equivalent (first equation).
2. according to the multiplication law of probabilities, the joint probability of the n events $(N_d > j), j = 1,\ldots, n$ factors as shown in the second line of eq. (5);

3. the third line of eq. (5) follows from the second one as $\left(\bigcap_{i=1}^{j}(N_f > i)\right) \Leftrightarrow (N_d > j)$.

   Eq. (4) and eq. (5) together finally yield

$$Pr\{N_f > n \mid \theta, \delta\} = \prod_{j=1}^{n}\frac{(1-\theta)}{1+(j-1)\delta} = \frac{(1-\theta)^n}{\prod_{j=1}^{n}(1+(j-1)\delta)} = \frac{(1-\theta)^n}{\prod_{j=1}^{n-1}(1+j\delta)} \tag{6}$$

In order to write down the likelihood function of the model we need also the probability that the first failure occurs at time n in the case that the values of the parameters are known. The event: "first failure at time n" can be intersected with the event "survival till time $n-1$" because the former implies the latter and the after the conjunction we obtain the same event. The effect of the conjunction is to make it easier to derive the probability we are interested in, as we can write

$$Pr\{N_f = n \mid \theta, \delta\} = Pr\{((N_f = n)\cap(N_f > n-1))\mid\theta, \delta\} =$$
$$= Pr\{N_f = n \mid N_f > n-1, \theta, \delta\}\cdot Pr\{N_f > n-1 \mid \theta, \delta\} = \tag{7}$$
$$= \frac{(1-\theta)^{n-1}}{\prod_{j=1}^{n-2}(1+j\delta)}\cdot\frac{\theta + (n-1)\delta}{1+(n-1)\delta} = \frac{(1-\theta)^{n-1}[\theta+(n-1)\delta]}{\prod_{j=1}^{n-1}(1+j\delta)}$$

Indeed line 2 of eq (7) is obtained from line 1 by the multiplication law of probabilities; line 3 is obtained from line 2 by taking into account eq. (3) and eq. (6).

2.1.2 *Generalization of the model to field data*

The most general type of data one can observe relatively to components which eventually fail on demand is as follows
- k_i components survived up to the n_i-th demand, $i=1,...,s$;
- r components failed, the failures occurred at the demands $n_{s+1},...,n_{s+r}$.

Eqs (6) and (7) supply us with the probabilities of all the elements of the above sample D whose likelihood function $\mathcal{L}(D|\theta,\delta)$ can be constructed by suitably combining those equations, this yields:

$$\mathcal{L}(D|\theta,\delta) = \prod_{i=1}^{s}\left(\frac{(1-\theta)^{n_i}}{\prod_{j=1}^{n_i-1}(1+j\delta)}\right)^{k_i} \cdot \prod_{i=s+1}^{s+r}\frac{(1-\theta)^{n_i-1}[\theta+(n_i-1)\delta]}{\prod_{j=1}^{n_i-1}(1+j\delta)} =$$

$$= \frac{(1-\theta)^{\alpha}\cdot\prod_{i=s+1}^{s+r}[\theta+(n_i-1)\delta]}{\left(\prod_{i=1}^{s}\left(\prod_{i=1}^{n_i-1}(1+j\delta)\right)^{k_i}\right)\cdot\prod_{i=s+1}^{s+r}\prod_{j=1}^{n_i-1}(1+j\delta)} \quad ; \tag{8}$$

where $\alpha = \sum_{i=1}^{s} n_i \cdot k_i + \sum_{i=s+1}^{s+r} n_i - r$

2.2 *Assessing the prior*

One difficulty could appear when combining the joint PDF of the Prior knowledge of the analyst with the Likelihood (field data). This problem always exists but it is much simpler if a conjugate family of PRIOR is available in choosing PDFs belonging to the same mathematic family (Conjugate Distributions).

Unfortunately the probability model derived in the previous section does not belong to the exponential family, and then no class of conjugate priors on the unknown parameters is available (De Groot, 1970).

The use of conjugate priors, when these exists, enables the analyst to quantify in a mathematically rigorous way (Cifarelli, 1991) the relative impact of the prior knowledge on the assessment of reliability of the component of interest (Posterior Assessment of the reliability).

The very advantage of the Bayes approach with respect to the frequentist one is the possibility of keeping under control the relative weight between data and prior knowledge. Indeed also frequentists use prior knowledge (Clarotti, 1993).But denying the use of prior knowledge makes the classical statistics incapable of ascertaining the reasonableness of the results which are eventually arrived at.

In the Bayes frame, one has always the possibility of governing the interaction of prior knowledge and data as, even when a family of conjugate priors does not exist, one can always use heuristics for making sure that the impact of prior knowledge on the final results is not too large.

For what concerns the case at hand, we note that the sub-sample formed from the observed performances of the components on the first demand can be considered as a sample from a Bernoulli distribution. Indeed on the first demand, as it can be verified by setting $n=1$ in eqs. (3) and (4), the performances of the components depend on just θ. The Bernoulli model belongs to the exponential family and then it has a class of conjugate priors(De Groot, 1970). Yet, the prior on θ and δ can be factored according to eq. (9)

$$\pi_0(\theta,\delta) = \pi_0(\theta)\cdot\pi_0^1(\delta|\theta) \tag{9}$$

We can then proceed as follows for what concerns the assessment of $\pi_0(\theta, \delta)$:

i. we choose $\pi_0(\theta)$ among the priors conjugate to the Bernoulli model (Beta distribution) in a way that the posterior on θ obtained by observing just the performances of the components on the first demand, "will be shaped mainly by data" (i.e. the role of the prior knowledge in shaping will be modest) insofar as this is compatible with "reasonable" guesses of the unreliability of a fresh component";

ii. we choose $\pi_0^1(\delta|\theta)$ among the uniform priors to mean that, for any value of θ, no range of values of δ is a priori preferred except, maybe, those ranges which are suggested by sensible engineering judgment.

2.2.1 Assessing $\pi_0(\theta)$

The mathematical expression of a probability density function (pdf) which belongs to the conjugate family of the Bernoulli model (this family is referred to as a two parameters beta family) is as follows

$$\pi_0(\theta) = \frac{\Gamma(a)}{\Gamma(aq_0)\Gamma(a(1-q_0))} \theta^{aq_0-1}(1-\theta)^{a(1-q_0)-1} \tag{10}$$

where Γ is the gamma function.

The pdf's of the above family possess the following properties (Cifarelli, 1991).

The hyperparameter q_0 is the analyst's a priori statement of the probability that the Bernoulli process being studied takes the value 1. In our case, q_0 is our a priori statement of the probability that the component will fail on the first demand (a priori statement of the unreliability of a fresh component).

In the case of Bernoulli processes (binary processes) the sufficient statistics for the unknown parameter θ (the unreliability of a fresh component in our case) is the arithmetic mean of the 1's and 0's we observed. This means, loosely speaking and with reference to the performance of the aging component on the first demand, that the "best" estimate of the unreliability of the fresh component we can obtain from the observed data is the ratio $\bar{x}$ between the number of components which failed on the first demand and the total number of the components we tested.

The other hyperparameter which appears in eq. (10), i.e. the hyperparameter a, is such that: after observing m realizations of the Bernoulli process (m fresh components) the analyst's assessment of the unknown parameter θ will be

$$\bar{\theta}_n = \frac{a}{a+m}q_0 + \frac{m}{a+m}\bar{x} \tag{11}$$

where $\bar{x}$ is the arithmetic mean of the 1's and 0's the analyst observed.

Since $\bar{x}$ is the value the data "suggest" for the unknown parameter θ, eq. (11) reads: after m observations the assessment of the unknown parameter will be due to data to an extent of $\frac{m \cdot 100}{a+m}\%$ and to the analyst's prior guess to an extent of $\frac{a \cdot 100}{a+m}\%$.

In the case of the aging component: if we observed some tens of components (say 60) and we choose for a a value one order of magnitude lower than the order of magnitude of the number of the on test components (say we choose the value 6), then the final assessment of the unreliability of a fresh component will be due to our prior guess to an extent of roughly 10%. Of course the lower the value of a the milder the impact of the prior guess on the final assessment will be but, as we shall see in a while, it is not "reasonable" to move from the value of m farther than one order of magnitude.

The expectation of a pdf such as those defined by eq. (10) is equal to q_0. The standard deviation σ of a pdf belonging to the beta family satisfies eq. (12)

$$\frac{\sigma}{q_0} = \frac{\sqrt{1-q_0}}{q_0} \sqrt{\frac{1}{\left(\frac{a+1}{q_0}\right)}} \tag{12}$$

The ratio $\frac{\sigma}{q_0}$ quantifies the spread of the pdf about its mean in a way that:

– values of a much smaller than $\frac{1}{q_0}$ would be "unreasonable" as they would lead to extremely flat priors reflecting the incapability of the analyst of specifying even just the order of magnitude of the unreliability of a fresh component;

– values of a much larger than $\frac{1}{q_0}$ would cause the prior to be sharply bell shaped which would preclude to the data the possibility of being the "main designer" of the posterior on θ. In order to implement the strategy sketched under task i. it must then be

$$q_0 \cdot a \approx 1 \tag{13}$$

In the light of eq. (13), very low values of a entail very large values of q_0 which means that a priori the analyst thinks that the fresh components are very unreliable and this is not "a reasonable guess". This consideration explains why the value of the hyperparameter a must not be much smaller than m.

2.2.2 Assessing $\pi_0^1(\delta|\theta)$

Since we want $\pi_0^1(\delta|\theta)$ to be uniform, we have just to specify, for any value of θ, the upper bound of the domain of variation of δ. If, on the basis of engineering judgment, we feel that after any demand the unreliability of the component cannot increase by an amount larger than the unreliability of a fresh component, then the upper bound of the domain of variation of δ corresponding to a given value of θ, is simply value of θ itself. A slightly more general model is the one which assesses the upper bound to be $\gamma \cdot \theta$. The idea of an upper bound proportional to the value of θ seems to represent well the case of an aging phenomenon which is sensible but not catastrophic and it has been assumed for the study which is reported in section 3. In general, the problem of selecting the value of γ is solved via sensitivity analyses that is: reliability calculations are carried out over and over again with different values of γ to make sure that the conclusions one arrives at are "stable" with respect to the variation of the parameter value. This could not be the case and then one should recognize that data are not enough to support any "linear aging model" or even to support any aging model whatsoever. In this case, the assessment of the aging phenomenon would be completely at the mercy of the engineering judgment (this would not be due to the use of Bayesian methods but simply to the lack of data).

2.3 Settling the numerical integration problems

After devising a probability model and assessing a prior, the Bayes theorem permits us to obtain the estimates of interest to safety engineers namely: component reliability, the expected values of the unknown parameters etc.

Suppose we are interested to the probability that the aging component will survive till time $n_1 + n_2$ given that it survived till time n_1, we are then to compute the following integral

$$P(N_f > n_1 + n_2 | N_f > n_1) = \frac{\displaystyle\int_0^1 d\theta \int_0^{\gamma\theta} d\delta \cdot \frac{(1-\theta)^{n_1+n_2}}{\displaystyle\prod_{j=1}^{n_1+n_2-1}(1+j\delta)} \mathcal{L}(D|\theta, \delta)\pi_0(\theta, \delta)}{\displaystyle\int_0^1 d\theta \int_0^{\gamma\theta} d\delta \cdot \frac{(1-\theta)^{n_1}}{\displaystyle\prod_{j=1}^{n_1-1}(1+j\delta)} \mathcal{L}(D|\theta, \delta)\pi_0(\theta, \delta)} \qquad (14)$$

Of course, given the involved analytical expressions of the functions which appear in the above integrals the latter must be computed numerically but this is not too difficult a problem. Indeed recently developed numerical integration techniques (Clarotti, Procaccia & Villain 1996) permit to compute rapidly the integrals which arise during the Bayesian analyses of data from the aging distributions.

3 APPLICATION TO THE OPTIMIZATION OF THE MAINTENANCE OF DIESEL ENGINE LININGS

3.1 *The problem*

In any EDF NPP the auxiliary electric power is supplied with 2 * 5 MW Diesel Generators, insuring a 100% redundancy of the necessary energy, in case of incident on the plant.

The aging of certain subcomponents of the Diesel engine, such as the linings (20 per Diesel set) is essentially governed by the number of annual starts up.

To prevent this aging, the basic present maintenance policy of linings is the following:
- each lining undergoes one endoscopic examination every year during the plant refueling. Evidence of degradation (cracks, scratches...) induces systematic conditional maintenance,
- every 5 plant operating years all the linings are replaced (preventive maintenance). For the most part of plant operators the linings are still in a good operability state after 5 years of operation (about 125 starts up and 120 operating hours in average.)

It has been asked to determine the best maintenance policy of the Diesel linings.

3.2 *Dieselbayes software*

Bayesian Statistical Decision Theory (Procaccia, Piepszownik, Clarotti) has been used to optimize the corrective (what to do if cracks are observed during the annual endoscopic examination), and the systematic preventive maintenance (what is the best periodicity of lining replacement taking into account the probability of aging failure, and the consequences of the failure).

We shall only speak, in the following, of the second module (preventive maintenance) of a specific RCM (Reliability Centred Maintenance) software which has been developed for Diesel maintenance optimization : Dieselbayes.

The software builds Decision Trees for different scenarii to be tested (example, Figure 1), such as preventive maintenance periodicities, computes the risks for the Diesel set to be called for a mission, the number of starts up for test, the corresponding risk of failure, and the economical impact of the failure. All the feedback experience data (5 failures, 13 degradations, 31715 demands, 479 censured data, the reliability of the grid...), and the economical data (cost of one replacement $C_1=C$, cost of elementary repair, Unavailability cost..) are included in the software and can be updated.

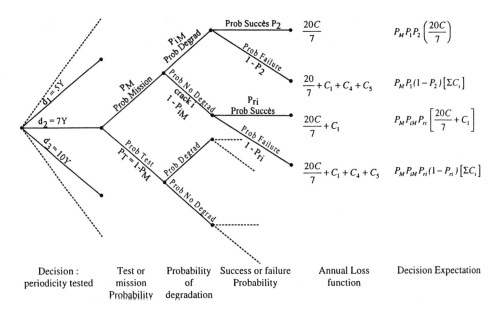

| Decision : periodicity tested | Test or mission Probability | Probability of degradation | Success or failure Probability | Annual Loss function | Decision Expectation |

Figure 1. Decision Tree.

The aging risk of failure is computed with the discrete Bayesian model given above, and can be approximated by an equivalent Weibull model in the PC version of the software.

One example of sensitivity analysis of aging calculation with the discrete Bayesian model is given in the Table 1.

Table 1. Sensitivity Analysis of the model (n_1 = 171 demands; n_2 = 1 demand)

Gamma (aging rate)	Unavailability U (n_2/n_1)	θ distribution		δ distribution	
		$\bar{\theta}$	σ_θ	$\bar{\delta}$	σ_δ
0.10	1.75 E-3	1.8 E-4	8.2 E-5	9.3 E-6	2.7 E-6
0.50	2.03 E-3	8.6 E-5	7.1 E-5	1.15 E-5	3.1 E-6
1.00	2.08 E-3	6.6 E-5	6.8 E-5	1.2 E-5	3.1 E-6

The more coherent result between data and Prior corresponds to gamma = 1, $\bar{\theta}$ is the probability of failure at the first demand of the fresh component. The aging is important because the conditional probability of failure at 172 demands is 2.1 E-3/d (compared to 6.6 E-5/d).

Applicating these results in the Decision Tree it is found that the optimal preventive maintenance periodicity is about 10 years (Figure 2).

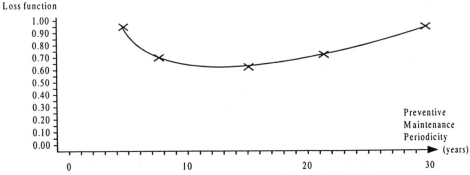

Figure 2. Determination of the optimal periodicity of Preventive Maintenance of Diesel Lining.

4 CONCLUSION

It has been shown in this paper that it is possible to modelize discrete aging of equipments (aging on demand) in using Bayesian inference.

This modeling has been used in a Decision Tree to determine the optimal preventive maintenance periodicity of Diesel engine linings.

The result is an extension of this periodicity from 5 years to 10 years, with practically the same limited risk, but with a substantial economical gain when regarding a population of 120 Diesel Generators.

REFERENCES

Cifarelli, M. 1991. L'approccio predittivo dell'inferenza. *In W d i stati. Racugno (ed), Problemi stica Bayesiana*: 49-77. Cagliari: SIS (in Italian).

Clarotti, C. A. 1993. Decision making. Drawbacks of the classical statistics approach. In R. E. Barlow, C. A. Clarotti, & F. Spizzichino (eds), *Reliability and decision making* : 1-12. London: Chapman & Hall.

Clarotti, C. A., Lannoy, A., & Procaccia, H. 1996. Modélisation de la fiabilité d'un matériel soumis à des sollicitations. *EDF Report HP-20/96/015/A* (in French).

Clarotti, C. A., Procaccia, H., & Villain, B. 1996. Bayesian analysis of general failure data from an ageing distribution: advances in numerical methods. *In C. Cacciabue & I. A. Papazoglou (eds)*, Probabilistic safety assessment and management: 1289-1294. Heidelberg. Springer & Verlag.

De Groot, M. H. 1970. Optimal Statistical Decisions. New York. *Mc Graw Hill* (chapt 9).

Procaccia, H., Cordier, R., Muller, S. 1996. Application of Bayesian Statistical Decision Theory for a Maintenance Optimization Problem .*Reliability Engineering & System safety.*

Procaccia, H., Piepszownik, P., Clarotti, C. 1992. Fiabilité des équipements et Théorie de la Décision Fréquentielle et Bayesienne. *Editions Eyrolles, Collection de la Direction des Etudes et Recherches d'EDF N° 81.*

Reliability analysis and safety evaluation on a nuclear power plant

Th. Meslin
Electricité de France, Saint Laurent, France

ABSTRACT: In many developed countries, Nuclear Power Plant have been built in the 60's and 70's, especially after the oil crisis. These plants are now often more than 10 or 15 years old and periodic safety re-evaluations are requested to demonstrate their ability to stay in operation with a satisfactory safety level.

Among the different aspects of the safety re-evaluation, the analysis of the technical behaviour of the main safety-related components is necessary to demonstrate the conformity with the design hypothesis.

In this context, the French nuclear Power plant of St-Laurent-des-Eaux developed tools in order to monitor the reliability parameters of the safety related components (failure rates per demand, hourly failure rates, mean times to repair, unavailability rates ....). These parameters are calculated on a yearly basis using the Bayesian statistical method.

The study of the evaluation of these parameters, since the start up of the plant in the early 80's, is a useful tool to help in identifying the ageing phenomena and validating the maintenance policies.

The paper describes the method used to monitor the reliability of the components and gives some examples of interesting behaviours.

The main safety related components are taken into account (more than 2,500 on each unit, grouped in one hundred families). Twelve years after the start up, no ageing phenomena involving any safety problem can be shown.

In some cases, the positive effects of design modifications, or operating practices (including preventive maintenance rules) are put in evidence.

We have also found in some other cases, generic or typical failures are associated with the same kind of component.

These reliability analyses are still under development and must be extended to the periodical test analysis and maintenance optimisation ; but the actual results are still encouraging.

Field application of 'Augur' ultrasonic system during RBMK NPP Unit ISI and its impact on pressure boundary integrity

A. I. Arzhaev & V. A. Kiselyov – *ECS MAE RDIPE, Moscow, Russia*

V. G. Badalyan & A. Kh. Vopilkin – *Echo+, Moscow, Russia*

B. P. Strelkov – *RDIPE, Moscow, Russia*

V. N. Vanukov – *RF Minatom, Moscow, Russia*

V. A. Aladinsky & V. O. Makhanev – *SRC NPO TSNIITMASH, Moscow, Russia*

ABSTRACT: Recent developments in Russia in the areas of ultrasonic testing and regulatory assessment of In-Service Inspection (ISI) results are discussed taking into account first results of their practical on-site application.

1 INTRODUCTION

Insurance in pressure boundary integrity of NPP Unit is strongly influenced by technical capabilities and efficiency of metal condition monitoring system and ISI as its main part. Ordinary ultrasonic testing tools and procedures have limitations in their flaw sizing and positioning capabilities as applied to piping welds. Major problems arise for welds and repair zones of welds made by welding materials of austenitic type. Problems of appropriate procedures and methods development for regulatory assessment of ISI results with respect to the level of data reliability also are very important. Some steps towards modification of scheme for ISI results assessment are presented herebelow. Results of practical modern UT tools on-site application during 1996 year outage at RBMK NPP Unit are also discussed.

2 «AUGUR 4.2» MAIN CHARACTERISTIC FEATURES

In order to correspond today requirements for NPP equipment and piping strength reliability process of permanent enhancement of ISI tools & procedures is under way. Nowadays on of the most perspective area of such R&D is development of a new generation computerized ISI systems. These systems in parallel with capabilities of ordinary UT techniques allow to reconstruct with high resolution level the image of object inner volume and increase the ISI gathered amount of information.

Recently developed in Russia in the framework of RF Minatom program series of computerized defectoscopes "Augur" for inspected area image reconstruction belong to these class of UT devices. One of them is device with coherent data processing for expert ISI named «Augur 4.2». In the end of year 1995 it have passed appropriate testing under supervision of Russia regulatory authorities and was recommended by them to be applied on-site during expert ISI. One of the main characteristic feature of these device and procedure is the capability of precise flaw image reconstruction which enables expert to discover flaw real geometric parameters. Capabilities of good revealing planar type defects make «Augur 4.2» an advantageous tool for ISI performance.

Defectoscope «Augur» is a complex device including Notebook PC connected to electronic unit and mechanical 2D scanner. Expert judgement on flaw sizes is obtained by the means of «Augur 4.2» ISI records processing. Based on this information strength and lifetime assessment could be updated. Inspection records obtained with the help of «Augur» are stored for ten years in full amount and could be retrieved and reanalysed, for example, for the purpose of thorough comparison with ISI records obtained during later inspections.

Validation of «Augur 4.2» according to internationally recognized procedure has been performed in the year of 1996 in Risley NDE Validation Center (AEA Technology, UK). According to the program of validation tests there were performed inspections of a set of certified samples with different type flaws.

The main purpose of expert ISI performance at RBMK NPP Unit primary circuit piping welds was «Augur 4.2» application to zones where some flaws had been revealed by previous scheduled manual UT inspections. This information was valuable for decision-making about further surveillance of these zones or corrective maintenance. Before inspection performance calibration test was carried out using test-blocks with artificial flaw models. Inspection has been performed with the use of shear-wave transducers with main frequency 2.5 MHz and inclination angle of 42 and 62 degrees.

Expert ISI was performed by means of «Augur-4.2» on MCC piping weld zones of RBMK NPP Unit where some indications had been revealed by previous manual UT inspections or repair had taken place. Measurements of two types have been performed in compliance with procedure on "Augur" application which provided high level of resolution of flaw images both along and transverse weld. Measurements have been performed from both sides of weld (where available) to obtain maximum information. «Augur» records were rewritten from acquisition computer to special temporary storage device of high information capacity (ZIP-diskette) and then have been processed on another computer outside contaminated zone. During further processing of records flaw images were restored, analysed and described in protocol of expert ISI in such parameters as length, upper and lower edges distance from outer surface of the pipe. Primary image of one of flaws discovered by «Augur» is shown on Figure 1.

3 ROLE OF EXPERT ISI IN THE GENERAL SCHEME
OF ISI RESULTS ASSESSMENT

The set of regulatory documents in Russia nuclear industry had no analogue for ASME Boiler and Pressure Vessel Code (ASME 1995) as a unified document for the stage of NPP equipment and piping operation. Basic documents for NPP Unit design stage (such as GAN 11986, 989, 1990) are valid for the stage of operation also and there is set of special regulatory procedures referring to some specific aspects as ISI typical programs, flaw acceptability assessment, etc.

In 1996 general scheme for ISI performance and its results assessment was in explicit form introduced into the last revision of typical ISI program for RBMK-1000 equipment and piping of Leningrad NPP Units (LNPP 1996). This was a result of both technical specialist of RDIPE and their colleagues from one side and leading experts of central office of regulatory authority form another side. This scheme is shown in Figures 2 and 3 bellow. Scheduled ISI is to discover some inhomogeneities in inspected weld areas and expert ISI is used to reveal the flaw details.

One of the most advantageous expert ISI means have become «Augur 4.2» tool and procedure also adopted by Gosatomnadzor of Russia and included into 1996 revision of typical ISI procedure mentioned above. One can see from Figure 3 that advantages of fracture mechanics application to flaw acceptability assessment are available only if means of expert ISI give real flaw dimensions. This provide reasonable level of conservatism in assessment, possibility to assume plant and weld specific features. Scheme of such analysis performance and criteria for flaw acceptability assessment are given in the regulatory procedures (RDIPE & VNIIAES 1988, 1991; RDIPE & Gidropress 1993).

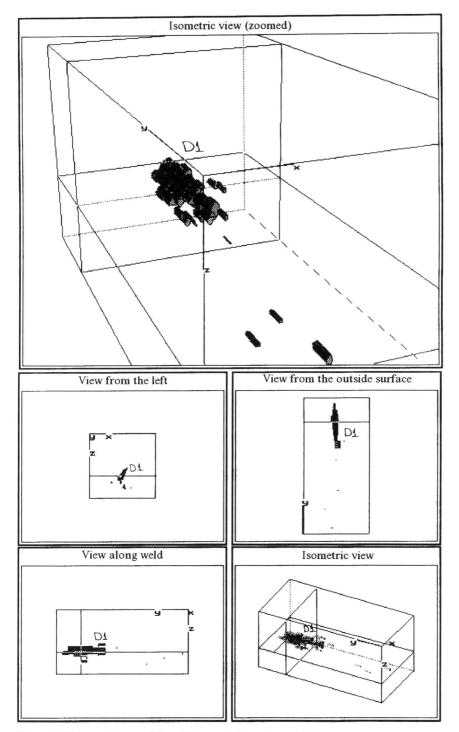

Figure 1. Primary image of flaw D1 revealed by «Augur 4.2».

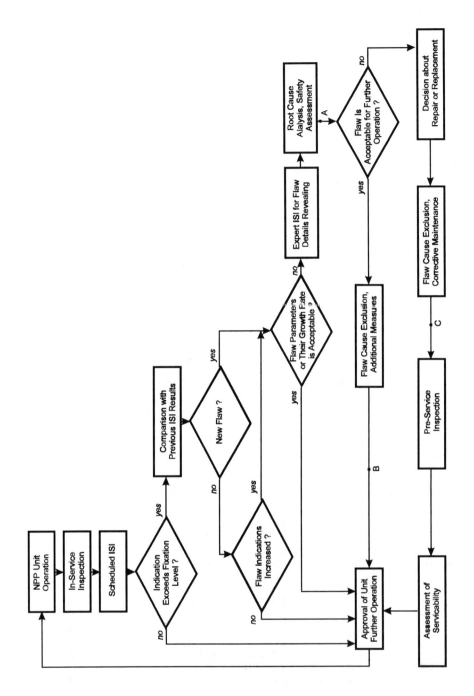

Figure 2. General Scheme of ISI Rsults Assessment

100

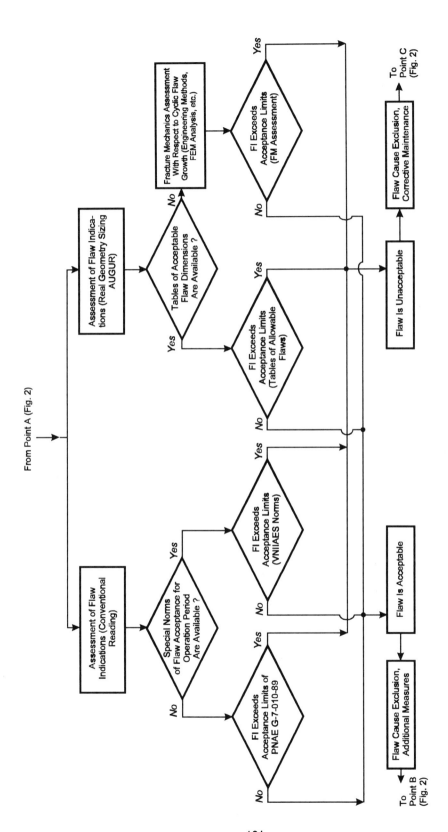

Figure 3. Logic Scheme for Flaw Acceptability Assessment On the Basis of Expert ISI Results (FI - Flaw Indication)

4 MAIN RESULTS OF «AUGUR» FIELD APPLICATION WITH RESPECT TO PRESSURE BOUNDARY INTEGRITY

In brief, the results of «Augur 4.2» ISI for Main Circulation Circuit piping weldments at RBMK NPP Unit are discussed. The representative set of weldments were examined: site welds (with pearlitic type weld metal) and two types of site weld repair zones with different welding technologies: using Cr-Ni filler metals and Ni-based alloy. Results of «Augur 4.2» inspection were compared with manual ultrasonic inspection records in terms of defect detection, characterization, positioning and sizing: in order to estimate correlation between these data (Figure 4).

Comparative analysis for both types repair weldment zones showed the following:

1) All defects detected by manual inspection were detected by «Augur 4.2» system too. But «Augur 4.2» indicated (by upper edge of defect) that the defects were located closer to the internal surface of pipe in comparison with manual control data.

2) «Augur 4.2» provided necessary data for precise fracture analysis as applied to several near-by-located defects.

3) There were several complex geometry cases for which one defect was detected by manual control, but two additional defects located at the same distance from outer surface, but in parallel planes were detected by «Augur».

4) Using the results of the identification analysis recommendations on «Augur 4.2» application on the basis of manual UT inspection records were formulated: first of all, for defects with indications less than 28-30 mm, or if group of near-located defects was detected.

During previous ISI performed by manual UT no growth of flaw indication parameters were revealed. But there were difficulties with flaw characterization: according to appropriate manual UT procedure they have been treated as «reflectors». On the basis of «Augur 4.2» records real position, type and sizes of these flaws have been discovered. On the basis of this information it was possible to characterize flaw as independent or take into account their interaction. Flaw dimensions in the direction of wall thickness in major cases were less than 2 mm. Finite element analysis showed that these defects likely had originated after first passes of repair welding and are caused by high level residual tensile stresses. Location of these flaws near the boundaries of cut-off zone together with information mentioned above made it possible to make a decision that they had likely technological cause.

Due to the fact of small flaw dimensions discovered their possible growth to critical sizes can be surely excluded. Nevertheless surveillance of these flawed zones by means of «Augur» is foreseen during further ISI to reveal flaw dimensions growth.

As applied to site pearlitic welds the following conclusions have been done:

1) All defects detected by manual control were detected by the «Augur 4.2» system too. As a rule, good correlation was observed between «Augur» coordinate of defect upper edge and manual UT inspection records.

2) Correlation between amplitude or equivalent square data of manual control and defect height showed by «Augur» were poor.

3) There were cases in which the extensive defect, detected by manual control, was cut into two or three isolated defects with small length by «Augur».

4) As applied to long defects detected in the middle part of wall thickness «Augur» control data had no effect on the defect acceptance. The rejecting minimum length and wall coordinate obtained by manual control were determined.

Using «Augur 4.2» data on defect configuration the strength and fracture mechanics estimations were carried out in accordance with defect assessment regulatory procedure (RDIPE & VNIIAES 1991). A number of defects with manual control indications clarified by «Augur 4.2» records have been accepted for further operation with prescription of next year «Augur» inspection. In order to be most effective in practical application «Augur 4.2» examination procedure should take into consideration the recommendations of defect assessment method, which define guaranteed parameters of «safe» and «need to be repaired» defects.

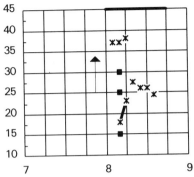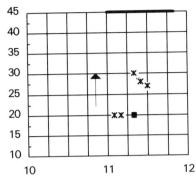

Figure 4. «Augur 4.2» (crosses) and manual (solid squares) repair weld examination data presented in 2D coordinates (horizontal axis — hoop position in hours, vertical axis — depth from the outer surface of the pipe in mm): solid line — area of «Augur 4.2» examination, arrow — repair depth. Left: manual control showed three defects one under another, «Augur» showed three independent defects. Right: «Augur» detected new defect with the height 1.5 mm near the bottom of the repaired zone.

The maximum benefit of «Augur 4.2» application is achieved if defect acceptability analysis incorporates results of computer finite element modelling (see Figure 3) and data of material mechanical properties and load monitoring.

5 CONCLUSIONS

Material presented hereabove makes it possible to draw the conclusion as follows:

1) effective tool for precise UT inspection of welds and base metal of NPP equipment and piping have been developed, validated, approved by Russian Regulatory Authority for application at NPPs in RF and successfully used as a mean of expert ISI at RBMK NPP Unit in 1996;

2) application of «Augur-4.2» to different type of weld zones have showed it practical capabilities in revealing real dimensions of flaws both in pearlitic and austenitic type weld materials;

3) general scheme for ISI results assessment have been included in explicit form into set of regulatory documents in Russia nuclear industry clarifying conditions of fracture mechanics methods application for flaw acceptability assessment;

4) comparison of manual UT data and «Augur-4.2» records have shown their good compliance in general and increased amount of information about inspected zones revealed by means of «Augur-4.2» which helps to reduce the level of ISI results uncertainty and therefore give additional insurance of pressure boundary integrity.

REFERENCES

ASME 1995. ASME Boiler and Pressure Vessel Code. Section XI. Rules for Inservice Inspection of Nuclear Power Plant Components.
GAN 1986. Equipment and Piping Strength Assessment Regulations. PNAE G-7-002-86. Moscow: Energoatomizdat (in Russian).
GAN 1989. Rules for design and safe operation of NPP equipment and pipes. PNAE G-7-008-89. Moscow: Energoatomizdat (in Russian).
GAN 1990. NPP equipment and piping. Welding and cladding. Examination rules. PNAE G-7-010-89. Moscow: Energoatomizdat (in Russian).

LNPP 1996. Typical ISI program for base metal and weldments of equipment and piping of systems important to safety of Leningrad NPP units with RBMK-1000 reactors. St-Petersburg-Moscow.

RDIPE & VNIIAES 1988. Method for determination of allowable flaw sizes in equipment and pipe metal during NPP operation (M-01-88). Moscow (in Russian).

RDIPE & VNIIAES 1991. Procedure for allowable flaw sizes evaluation in metal of equipment and piping during NPP operation (M-02-91). Moscow (in Russian).

RDIPE & Gidropress 1993. Assessment of NPP piping according to leak-before-break criteria (M-TPR-01-93). Moscow (in Russian).

Damage mechanics applications

Ageing of Materials and Methods for the Assessment of Lifetimes of Engineering Plant, Penny (ed.)
© *1997 Balkema, Rotterdam, ISBN 90 5410 874 6*

The thermomechanical material state and integrity retaining of reactor vessel under an anticipated accident

V. L. Danilov, M. V. Dobrov & S. V. Zarubin
Department of Applied Mechanics, Bauman Moscow State Technical University, Russia

Y. Fautrelle
Laboratory of Magnetic Melt Hydrodynamics for Application to the Metallurgy, National Polytechnic Institute of Grenoble, France

ABSTRACT: This paper discusses nuclear reactor vessel response under a severe accident, when the core is melted. During an accident the melt of core flows down to the bottom of reactor vessel forming corium. As a result, either the vessel bottom will be melted down or its rupture will occur on account of intensive high-temperature creep. Depending on the probabilistic proceeding of anticipated accident, different variants of reactor bottom break down have been obtained for the elliptical bottom of a water-water vessel type reactor. Analysis indicates possibility to retain reactor vessel integrity and hence to hold corium in a reactor under high cooling water level.

1 INTRODUCTION

A severe accident in a nuclear power plant became a matter of close attention and investigations in the 1980s and it is becoming more and more acute in connection with the new requirements for plant safety and new nuclear power plant project licensing. Apart from maximum design accidents there are anticipated accidents which cause more damage. To resist the former, safety systems and protection structures are being developed and constructed. The reactor becomes unprotected in the case of anticipated accidents, because they can be caused only by very unlikely events, the protection against which is not expedient, and by inadequate plant personnel actions too. This is the very type of accident at the Three-Mile-Island and Chernobyl plants.

The probabilistic development of events at severe accident due to coolant loss and inability of effective reactor core cooling and reactor vessel, as a whole, comes to the following. With temperature increase of fuel element claddings due to the decrease of heat removal the latter begins intensively swelling locally or in the entire volume and as a result it fails. Hereby, a simultaneous chemical reaction of zirconium alloy with vapor and the development of the failure along the entire fuel assembly is going on. Following this due to the intensive temperature growth, a melting of reactor core structure elements begins. The created melt, flowing down to the core bottom, increases in volume. All mass of melt material is accumulated in the lower part of the core and held with support plate. Then the support plate melts out and the resultant corium in the form of a solid-liquid mixture creates a bath at the bottom of reactor vessel. Being uncooled, the bath is melting down to a state of complete liquid. At the same time the vessel bottom starts melting. As a result, either the vessel bottom melts through in some place or it breaks down due to its deformation under the effect of internal pressure and temperature gradient.

The thermal and structural response of the lower head vessel during a severe nuclear accident has been studied by others, however, in these studies either simplified methods of stresses determination have been used (Dosanjh & Pilch 1991) or model of fracture front motion has not been included (Witt 1994), (Chavez & Rempe 1994). This paper discusses the aspect of probabilistic high-temperature creep rupture. However, it is considered that the vessel bottom can be melted partially from inside and thus the process of mechanical rupture is accelerated.

2 BASIC EQUATIONS OF THERMAL MODEL

The non-stationary heat conduction differential equation for continuum with internal heat sources is applied to determine temperature response in the reactor vessel - corium system in axisymmetric case. This equation is written in the following form

$$c_p(T)\rho(T)\frac{dT}{dt} = div[\lambda(T)gradT] + q_p + q_f,$$ (1)

where c_p is the heat capacity, ρ is the material density, λ is the thermal conductivity, q_p is the function characterizing three-dimensionally distributed heat sources due to phase changes, q_f is the function of heat release in the corium due to permanent processes of radioactive decay in core fuel elements. The function q_p is accepted directly proportional to material crystallization (or melting) rate and depends on material properties (particularly, on phase change heat and crystallization interval). To describe q_p quantitatively, the fraction of solid phase φ in a two-phase (solid-liquid) zone is introduced. Then q_p is expressed as

$$q_p = -\rho L \frac{d\varphi}{dt},$$ (2)

where L is the specific melting heat.

Melting and crystallization processes are considered in equilibrium statement (the diffusion process in an elementary material volume are assumed amply to follow during phase change). Hence solid fraction - temperature relation is known from equilibrium material state diagram. Heat transfer caused by convection and radiation are taken into account on the boundaries of the reactor vessel - corium system.

To obtain temperature fields evolution a standard finite element procedure is applied (Zienkiewicz 1978).

3 STRUCTURAL ANALYSIS MODEL

3.1 Basic relationships

The analysis algorithm of stress-strain behaviour of the reactor vessel lower part under the conditions of non-stationary thermal influence is constructed for axisymmetric case. Let's formulate state equations applied for finite element procedure solution.

Introduce the strain vector ε and stress vector σ

$$\varepsilon = \{\varepsilon_r, \varepsilon_\vartheta, \varepsilon_z, \gamma_{rz}\}^T, \qquad \sigma = \{\sigma_r, \sigma_\vartheta, \sigma_z, \tau_{rz}\}^T.$$ (3)

The full strain rate vector are represented as $\xi = \xi^e + \xi^c + \xi^t$. The elastic strain rate vector ξ^e and stress vector σ are connected by the matrix relationship

$$\xi^e = \frac{d}{dt}\left[\mathbf{D}^{-1}(T)\sigma\right],\qquad (4)$$

where $\mathbf{D}$ is elastic matrix, whose components depend on temperature. Assuming that the material is isotropic matrix $\mathbf{D}$ will be written as

$$\mathbf{D} = \frac{E}{(1+\nu)(1-2\nu)}\begin{bmatrix} 1-\nu & \nu & \nu & 0 \\ \nu & 1-\nu & \nu & 0 \\ \nu & \nu & 1-\nu & 0 \\ 0 & 0 & 0 & 0.5-\nu \end{bmatrix},\qquad (5)$$

The creep strain rate vector is determined by means of creep theory equations for complex stress state

$$\xi_{ij}^c = \frac{3}{2}\frac{\xi_e^c}{\sigma_e}s_{ij}.\qquad (6)$$

Here s_{ij} are stress deviator components, σ_e is effective stress to be equal

$$\sigma_e = \left(\frac{3}{2}s_{ij}s_{ij}\right)^{1/2}.\qquad (7)$$

Effective strain rate is taken as

$$\xi_e^c = Q(\sigma_e, T) = A\exp\left(-\frac{T_0}{T}\right)\sigma_e^{n(T)},\qquad (8)$$

where T is temperature, A, T_0 are experimental constants, n is experimental function of temperature. Since the isotropic material is considered, the temperature strain rate vector is expressed as

$$\xi^t = \left\{\xi_0^t, \xi_0^t, \xi_0^t, 0\right\}^T,\qquad (9)$$

where $\xi_0^t = \alpha_t(T)\dfrac{dT}{dt}$. Here α_t is linear expansion coefficient, t is time.

The material rupture and the rupture front propagation are described within dispersed damages accumulation concept (Kachanov 1974). Damage function is calculated in each point simultaneously with stresses and strains. The damage function equation is written by considering the linear damage summation principle

$$\omega = \int_0^t \frac{Q(\sigma^{eq}, T)}{\varepsilon_u(T)}dt,\qquad (10)$$

where ε_u is characterized as an ultimate strain under uniaxial tension, σ^{eq} is equivalent stress that is defined by the Sdobyrev's criterion (Rabotnov 1966)

$$\sigma^{eq} = \begin{cases} (\sigma_{max} + \sigma_e)/2, & \text{if } \sigma_{max} > 0, \\ \sigma_e, & \text{if } \sigma_{max} \le 0, \end{cases} \tag{11}$$

where σ_{max} is maximal principal stress. The function ω varies from 0 to 1, and $\omega=1$ is the creep rupture condition.

3.2 Numerical procedure

A powerful vessel lower part heating is the given problem distinction, that can cause vessel bottom displacement of the order of wall thickness due to intensive high-temperature material creep. Therefore numerical step algorithm is constructed taking into account both physical (material) and geometrical non-linearity.

Expressions for creep and temperature strain increment vectors in a time interval $\Delta t_n = t_{n+1} - t_n$, where Δt_n, t_n, t_{n+1} indicate step length, step beginning, and step end are written on the basis of explicit procedure

$$\Delta \varepsilon_n^c = \xi_n^c \Delta t_n, \qquad \Delta \varepsilon_n^t = \xi_n^t \Delta t_n. \tag{12}$$

Then elastic strain increment vector is

$$\Delta \varepsilon_n^e = \Delta \varepsilon_n - \Delta \varepsilon_n^c - \Delta \varepsilon_n^t, \tag{13}$$

where $\Delta \varepsilon_n$ is full strain increment. On other hand, using (4) $\Delta \varepsilon_n^e$ can be expressed as

$$\Delta \varepsilon_n^e = \mathbf{D}_{n+1}^{-1} \Delta \sigma_n + \Delta \mathbf{D}_n^{-1} \sigma_n, \tag{14}$$

where $\Delta \mathbf{D}_n^{-1} = \mathbf{D}_{n+1}^{-1} - \mathbf{D}_n^{-1}$. Hence step stress increment will be

$$\Delta \sigma_n = \mathbf{D}_{n+1} \left(\Delta \varepsilon_n^e - \Delta \mathbf{D}_n^{-1} \sigma_n \right) \tag{15}$$

or taking relationship (13)

$$\Delta \sigma_n = \mathbf{D}_{n+1} \left(\Delta \varepsilon_n - \Delta \varepsilon_n^c - \Delta \varepsilon_n^t - \Delta \mathbf{D}_n^{-1} \sigma_n \right). \tag{16}$$

This relation is used in a standard procedure of finite element analysis (Zienkiewicz 1978).

During each time step the innerstep iteration process is implemented to correct stress-strain state. The equilibrium equation satisfaction is as convergence criterion of iteration process (Zienkiewicz 1978).

$$\psi = \int \mathbf{B}^T \sigma \, dV + \mathbf{F} \to 0, \tag{17}$$

where $\mathbf{B}$ is nonlinear strain matrix, $\mathbf{F}$ is vector of equivalent nodal loads due to applied surface and thermal loads.

The break down and melting of the solid continuum brings its distinction in the standard procedure of a finite-element analysis. The damaged and melted-down element is to some extent loosing its bearing strength. Therefore it is necessary either to exclude it from consideration, which brings about the reconstruction of a finite-element grid or to reduce stresses in such elements to the values corresponding to the physical sense of the process.

In this work the second variant of this problem solution has been used. For this purpose, an innerstep iteration process has been constructed, that is the process occurring completely for each time moment after the rupture and melt-down of one or several elements and which imitates the "unloading" of such elements. The given process includes a correction of inelastic strains in these elements, that is actually introduces some dummy inelastic strains that allow to ensure in ruptured and melted elements a required stress state. Three cases of bearing strength loss are considered, namely:

1) damage function $\omega=1$ and $\sigma_0>0$, where $\sigma_0=\delta_{ij}\sigma_{ij}/3$, δ_{ij} is Kronecker symbol;
2) damage function $\omega=1$ and $\sigma_0<0$;
3) the finite element temperature T exceeds the solidus temperature of vessel material.

4 RESULTS

The computations have been made for the elliptical bottom of a water-water vessel-type reactor. An anticipated accident was considered when the core was completely melted. At initial instant of accident corium was assumed to be liquid homogeneous medium with temperature of 3000 K. Internal pressure was taken to be 5 MPa. The accident development have been analysed for various external cooling emergency conditions. The computational results are presented for three basic cases:

1) a forced cooling is not used and the vessel is an contact with air (Figure 1);
2) a cooling water level is low (Figure 2);
3) a cooling water level is in excess of corium level in a reactor (Figure 3).

Numerical values in these figures characterize temperature and damage function for marked zones.

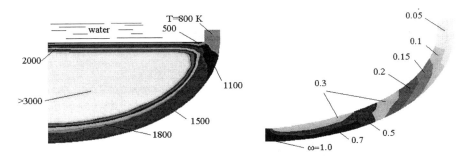

Figure 1. Temperature response of the corium-vessel system and damage distribution in the vessel bottom at instant of break-down (Case 1).

Depending on the probabilistic anticipated accident scenario, different variants of the reactor bottom rupture have been obtained: break down in the region of pole vicinity (Figure 1) and complete bottom break off causing the creation of a large core melted fragments release circuit (Figure 2). At first, in all cases damage accumulation took place in the vessel internal surface. It was on account of the sharp temperature increase of the boundary layers and as a consequence due to strong compression region appearance. The vessel bottom was partially melted in the most strongly heated zone for the both case 1 and case 2, that essentially accelerated damage accumulation in this zone and in adjoining regions. Creep began to dominate in external surface as vessel heat up. Therefore, damage level reached critical value faster there. After that rupture continued to spread in radial direction from external surface to internal surface until the complete break down. In case 3, thermal equilibrium was reached due to high water level and intensive creep did not be initiated. High water level ensures integrity retaining of reactor vessel that enables to hold corium in a reactor.

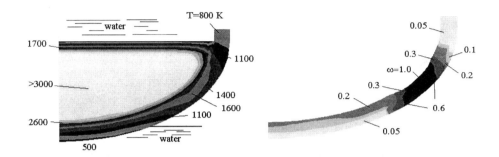

Figure 2. Temperature response of the corium-vessel system and damage distribution in the vessel bottom at instant of break-down (Case 2).

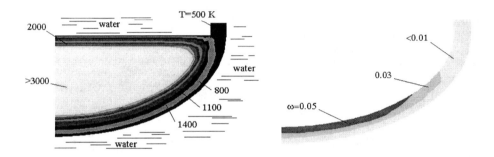

Figure 3. Temperature response of the corium-vessel system and damage distribution in the vessel bottom (Case 3).

The developed model allows to analyse the consequences of severe accidents in the tank nuclear reactor, and hence to determine trends of the design work in the reactor containment field, taking into account the possible evolution of an anticipated accident.

REFERENCE

Chavez, S.A. & Rempe, J.L. 1994. Finite element analyses of a BWR vessel and penetration under severe accident conditions. *Nuclear Engineering and Design.* 148: 413-435.
Dosanjh, S.S. & Pilch, M. 1991. Lower head creep-rupture sensitivity studies. *Nuclear Science and Engineering.* 108: 172-183.
Kachanov, L.M. 1974. *Fundamentals of Fracture Mechanics (in Russian).* Moscow : Nauka.
Rabotnov, Y.N. 1966. *Creep of structure elements (in Russian).* Moscow : Nauka.
Witt, R.J. 1994. Local creep rupture failure modes on a corium-loaded lower head. *Nuclear Engineering and Design.* 148: 385-411.
Zienkiewicz O.C. 1978. *The finite element method (3rd edition).* London: McGraw-Hill.

Ageing of Materials and Methods for the Assessment of Lifetimes of Engineering Plant, Penny (ed.)
© *1997 Balkema, Rotterdam, ISBN 90 5410 874 6*

Steel creep and creep rupture strength in environment containing hydrogen

V.L. Danilov & S.V. Zarubin
Department of Applied Mechanics, Bauman Moscow State Technical University, Russia

ABSTRACT: The phenomenological model of hydrogen steel embrittlement under long-duration tension stresses effect is constructed on the basis of dispersed damaged accumulation concept. The equation for damages accumulation rate describes various general non-elastic deformation and grain-boundary sliding influence upon microcracks formation process. The comparison of the calculated results and the data of long-duration creep and rupture experiment for low carbon steels in a water solution saturated with hydrogen.

1 INTRODUCTION

Cracks appearance in the oil-gas production equipment pipes and fittings is often connected with effect of hydrogen contained in oil products (Kushnarenko 1993). A metal surface accumulates adsorbed hydrogen, which partly volatiezes to environment. Other its part diffuses into metal (Alekseev et al. 1990). Negative hydrogen influence results in sharp metal plasticity decrease and decelerated rupture (Moroz L.S. 1984). Decelerated steel products rupture is the most dangerous in practice because in many cases it appears suddenly after long-duration stock-produced article operation under stresses being less then yield point.

As experiments show, under decelerated rupture cracks nucleating is impossible without preliminary microplastic deformation and such rupture often includes low-temperature creep elements even at a room temperature. For example, in tension test of smooth low carbon steel specimens in environment containing hydrogen at 20°C, creep was observed during all the holding time (to 180 days) of loaded specimens. Whereas in test in air, creep process died down rapidly (creep process creep strain rate reached zero about in 40-50 hours).

In this work an attempt is made on the basis at dispersed damages accumulation concept (Kachanov 1974) to construct the phenomenological model of hydrogen steel embrittlement under long-duration tension stresses effect. Let's note, rupture as a result of hydrogen embrittlement is many-stage process (Kolachev 1985). A description of such process requires consideration of all microcraks ensemble, their coupling with each other and different material structure defects, their timed evolution and merge in a main crack. It is unlikely, analytical rupture process description on such level is possible owing to its exceptional complexity. In this connection it is preferably to determine one or several rupture micromechanism playing dominating part in kinetics and evolution direction on the macrolevel of all damages ensemble.

2 PHENOMENOLOGICAL MODEL OF HYDROGEN STEEL EMBRITTLEMENT

Phenomenological relations are constructed taking account of simplified ideas about basic plastic deformation and rupture mechanisms. As a first approximation it is possible to adopt the following hydrogen cracking steel scheme. Hydrogen results in intercrystalline strength decrease (mechanism of this phenomen is not considered) and intensification of grain-boundary sliding. Its a result, favorable conditions for formation evolution and progressive defects accumulation on grain boundaries are created, what results in brittle intercrystalline rupture in the end.

Let's represent scalar damage parameter ω as the sum of two components

$$\omega = \omega_g + \omega_b .$$ (1)

For undamaged material ω is equal to zero, ω is equal to unity at instant of rupture. The augend ω_g is connected with damages accumulation process under normal conditions, the addend - with embrittling hydrogen influence. Let's write binominal expression for nonelastic strain rate too:

$$\xi^{(n)} = \zeta_g + \zeta_b .$$ (2)

The augend in this equation allows to describe a normal material deformation curve. The homographic expression is used for approximation ξ_g - stress relation:

$$\xi_g = \begin{cases} A_g \dfrac{\sigma - \sigma_g^*}{\sigma_Y - \sigma} , & \sigma > \sigma_g^* \\ 0 , & \sigma \leq \sigma_g^* \end{cases}$$ (3)

Parameter σ_g^* is actually a certain threshold stress, below of which creep is not become apparent. By means of introduction of additional kinetic equation for this parameter

$$\dot{\sigma}_g^* = B_g (\sigma_s - \sigma)^m \xi_g$$ (4)

it is possible to take creep hardening into account. Parameter σ_s is determined by material tension diagram under normal conditions

$$\sigma_s = f(\varepsilon_g) .$$ (5)

The addend in equation (2) characterize viscous flow along the grains boundaries due to intercrystalline strength reduction under hydrogen influence. Using traditional notation for power creep law and strain hardening, let's represent this component as follows

$$\xi_b = \begin{cases} A_b(c)\left[\sigma - \sigma_b^*(c)\right]^n (1 + B_b \varepsilon_b^k)^{-1}, & \sigma > \sigma_b^*, \\ 0, & \sigma \leq \sigma_b^*. \end{cases}$$ (6)

Coefficients A_b and σ_b^* equation (6) depend on hydrogen concentration c into metal.

To concretize kinetic equations for damage parameters, involved in equation (1), it was adopted, that damages accumulation process was immediately connected with metal deformation. For damages accumulation rate ω_g the equation was written as

$$\dot{\omega}_g = D_g(c)\xi_g \exp(-S\omega) . \qquad (7)$$

The addend in equation (1) is connected with defects accumulation as a result of intergrain sliding, therefore base factor having an effect on damages accumulation rate ω_b was considered to be strain rate ξ_g

$$\dot{\omega}_b = D_b\xi_b - g_R(\omega_b) . \qquad (8)$$

Possibility of damages "healing" during recovery after stresses removing is taken into account in equation (8). It $g_R=0$, initial metal properties recovery doesn't take place.

3 RESULTS

The mathematical model equations (1)-(8) are integrated comparatively easy and it simplifies constants determination in experimental data processing. Results of long-duration low-carbon steel (0.35% carbon) creep and rupture experiments under uniaxial tension in a water solution saturated with hydrogen (5%NaCl + 0,5%CH$_3$COOH) at 20°C are presented in the Figure 1-3.

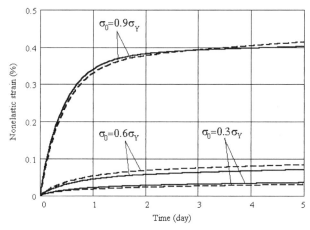

Figure 1. Nonelastic strain - holding time relations.

Smooth cylindrical specimens 6 mm across were held during time t under fixed load, after that, they were carried to rupture by tension with constant strain rate. At numerical computations of specimens deformation for the given loading regimes relative cross restriction ψ was determined in compliance with the following expression

$$\psi = 1 - \exp(-\varepsilon_0^{(n)}) , \qquad (9)$$

where $\varepsilon_0^{(n)}$ - nonelastic specimen strain at instant of rupture. At determination of coefficients in equations (3), (4), (6) and (8) "healing" was considered not to be and $g_R=0$. Experimental data are shown by solid lines, computational results - by dash lines in the all figures. The nonelastic strain - holding time relations under different stress levels σ_0 are presented in the Figure 1. The behaviors of $\varepsilon^{(n)}$ under stress $\sigma_0=0.9\sigma_Y$ (σ_Y - yield point) and more long-duration holdings are shown in the Figure 2. Results for ψ are referred to cross restriction for the given steel under tension in air similar results (ψ_0) are presented in the Figure 3.

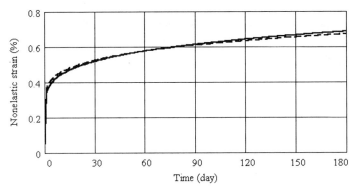

Figure 2. Nonelastic strain - holding time relations under stress $\sigma_0=0.9\sigma_Y$.

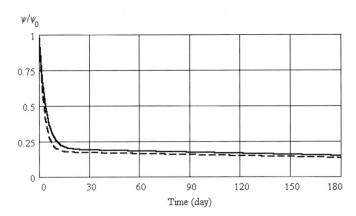

Figure 3. Relative cross restriction - holding time relations under stress $\sigma_0=0.9\sigma_Y$.

Performed theoretic computations being in compliance with experiment conditions showed adequacy of developed model of hydrogen creep and creep rupture strength. Let's note, suggested mathematical model allows quantitatively to describe known effect (Moroz L.S. 1984) of plasticity loss with deformation rate decrease for steels saturated with hydrogen.

REFERENCES

Alekseev, V.I., Kiselev, O.A. & Levina, I.V. 1990. High hydrogen pressure effect in phenomen of sulphide hydrogen corrosion cracking (in Russian). *Fiziko-Himicheskaya Mehanika Materialov.*. 26(2): 33-36.
Kachanov, L.M. 1974. *Fundamentals of Fracture Mechanics (in Russian).* Moscow : Nauka.
Kolachev B.A. 1985. *Hydrogen metal embrittlement (in Russian).* Moscow : Metallurgiya.
Kushnarenko, B.M. 1993. Strength of steels under sulphide hydrogen corrosion cracking (in Russian). *Zashita Metallov.* 29: 885-889.
Moroz L.S. 1984. *Mechanic and physics of material strains and failure* (in Russian). *Leningrad: Mashinistroenie.*

© 1997 Balkema, Rotterdam, ISBN 90 5410 874 6

Gradual failure of trusses in creep conditions

M.Chrzanowski & P.Latus
Department of Strength of Materials, Cracow University of Technology, Poland

ABSTRACT: The failure of structures in creep conditions is analysed by means of Continuum Damage Mechanics. Deterioration process induced by high temperature exposure and external loads leads to the total collapse of structures through several stages. These are: nucleation of macroscopic cracks, their propagation throughout structure's members, and formation of collapse mechanism. Step by step analysis (based on Finite Element Methods) allows for evaluation of characteristic times corresponding to the termination of each of the above stages. The ratios of these time values are used as a definition of safety margins. The effects of creep buckling and - consequently - development of damage zones in compressed bars is also considered. It is demonstrated that all this process is represented by so called "bathtub curve" indicating three stages of failure process: initial hardening, steady-state and final acceleration preceding structure's collapse.

The results of this work can be referred to the problems of safety evaluation for metallic structures working at elevated and high temperatures.

1 INTRODUCTION

Engineering structures composed of straight bars (or struts) joined by non-frictional hinges at which load is applied, bears name of trusses. It is obvious that this high idealisation of real structures which possess many imperfections, even if carefully made and assembled. Nevertheless, from analytical point of view this type of structures allows for clear identification and realisation of internal forces flow. These are also often used to carry pay-off loads e.g. in cranes or bridges. But being real they are very much influenced by above mentioned imperfections. These effects will be studied below for structures made of materials which in given environmental conditions exhibit time dependent phenomena: creep and material deterioration.

2 GRADUAL FAILURE OF TRUSSES

In trusses two types of struts can be distinguished: these under compression and under tension. It is well known (Ashby 1986) that at high temperatures metals exhibit progressive deterioration which may lead to delayed fracture. This phenomenon is governed mainly by tensile stresses, therefore truss struts under tension may rupture after some time of exposure to loads and high temperature. As stress distribution across struts under tension is uniform, the time when first macrocrack appears t_1 coincides with time when this strut ruptures at time t_2. But this occurrence does not exploit load bearing capacity of a truss. The process develops, and consecutive struts rupture until truss becomes kinematically unstable at time denoted as t_3. This process was already analysed by authors in [Bodnar et al. 1992, Bodnar et al. 1994] and lead to the observation that ratio of t_3 to t_1 can be significantly higher than one.

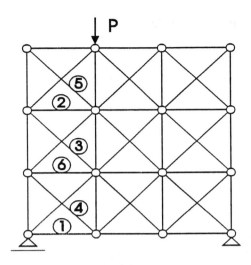

Figure 1. Multi-storey truss

The process of rupturing of consecutive struts undergoes well known „bathtub" effect [Le May 1995]. The results from [Bodnar et al. 1994] are used here to illustrate this effect for a multi-storey truss shown in Figure 1 (with bars labelled according to time sequence of failing). The plot showing the number of fractured bars referred to its rupture times is given in Figure 2. Assuming that this set of points can be interpolated by a smooth curve, and taking its first time derivative one can obtain a curve which represents the rate of structure's degradation (Figure 3). As in many technical processes three stages can be distinguished which correspond to decreasing, steady-state and increasing intensity of deterioration process.

3 GOVERNING EQUATIONS AND NUMERICAL PROCEDURES

The above results were obtained under assumption that material undergoes following constitutive equations.

For elasticity:

$$\varepsilon_{ij}^{e} = D_{ijkl}^{-1}\, \sigma_{kl} \tag{1}$$

For steady-state theory of non-stationary creep:

$$\frac{\partial \varepsilon_{ij}^{c}}{\partial t} = \gamma \left(\frac{\sigma_{e}}{1-\omega} \right)^{n-1} s_{ij} \tag{2}$$

For total strains:

$$\varepsilon_{ij} = \varepsilon_{ij}^{e} + \varepsilon_{ij}^{c} \tag{3}$$

where

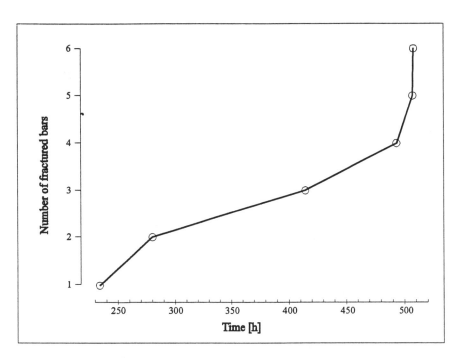

Figure 2. Time-sequence of fractured bars

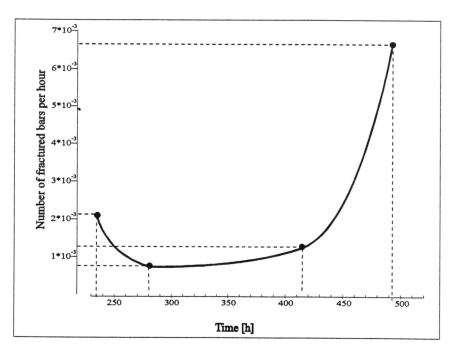

Figure 3. A bathtub curve for multi-storey truss

ε_{ij}^e and ε_{ij}^c are elastic and creep strain tensor components, respectively, of total strain ε_{ij}

s_{ij} - deviatoric part of stress tensor σ_{ij}

σ_e - is von Mises effective stress

D_{ijkl}, γ, n - are material constants,

and damage parameter ω is governed by following evolution law of classical damage theory by Kachanov-Rabotnov (Kachanov 1958):

$$\frac{\partial \omega}{\partial t} = A \left(\frac{\sigma_{eq}}{1-\omega} \right)^m \tag{4}$$

where equivalent stress σ_{eq} according to Hayhurst-Sdobyriev [Hayhurst 1972] is:

$$\sigma_{eq} = \alpha \sigma_1 + (1-\alpha)\sigma_e \tag{5}$$

and σ_1 is maximum principal stress, and A and m are material constants.

Critical value of damage corresponding to material failure is assumed to be equal 1.

Equation (2) demonstrates the effect of coupling between creep strains and damage, and therefore, it imposes stress redistribution.

In the case of uniaxial state of tress (which occurs in truss members) $\sigma_e = \sigma_1$. For struts under compression maximum principal stress $\sigma_1 = 0$, $\sigma_{eq} - 0$ and according to Equation (4) damage does not occur.

It is possible to use another theory [Lemaitre 1986] which accept damage growth under compression. Such behaviour was studied in [Bodnar & Chrzanowski 1990] but this phenomenon is not observed in most engineering materials. Much more important is that struts under compression may buckle, which leads to the unacceptable large lateral deformations. Then, structural members is subjected to compression and bending, and the latter may induce damage growth on tension side of a bar. This coupling between buckling and damage development was preliminary studied by authors in the paper [Chrzanowski & Latus 1996]. Here more detailed study will be presented covering a comparison of analysis with and without damage taken into account. This is especially important for *real* structures in which, due to unavoidable imperfections, bars under compression exhibit bending effect. For creeping structures it can be of prime importance, as lateral displacements start to growth immediately after loading, and keep on even if loading is maintained constant.

For metallic structures working in high temperature applications material characteristics are of nonlinear nature. Also, the effects of bending in compressed struts required a nonlinear geometry description to be used. Thus, no analytical solutions can be expected, and numerical procedures have to be applied. In this study a FEM system ABAQUS was used with user material module built, reflecting above constitutive equations (1-5). In all examples studied in this paper the following set of material constants was used: : n=3, m=2, $E=D_{1111}=$ 120 [GPa], $\gamma=1.3*10^{-6}$ [(MPa)$^{-n}$h^{-1}] , A=0.208*10^{-11} [(MPa)$^{-m}$ h^{-1}]

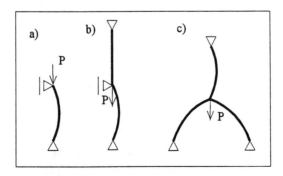

Figure 4. Simple trusses analysed in the paper

4 STRUT WITH IMPERFECTION

First, a single strut with end hinges under compressive load (Figure 4a) will be considered to evaluate the influence of initial imperfection and magnitude of loading force upon horizontal mid-span time dependent deflection and damage development. This imperfection , denoted as e, will be initial deflection in mid-span of a bar of length l.

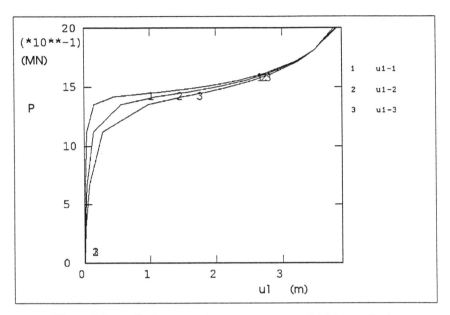

Figure 5. Force-displacement diagrams for different initial imperfections

Instantaneous elastic behaviour of strut is shown in Figure 5 for three different magnitudes of initial non-dimensional imperfection e/l = 0.001, 0.004 and 0.008, with l = 10 m. Corresponding lateral displacement u1 versus load P curves shown in this figure are labelled 1, 2 and 3, respectively. The Euler force for this bar is equal to 1.44 MN, and time dependent behaviour of this structure will be studied for loads below and above this critical force.

If material deterioration is not taken into account, but only time dependent lateral displacements are considered, the structure exhibits behaviour shown in Figure 6 for loading below critical one (P=1 MN), and in Figure 7 for loading above critical force (P=1.5 MN). A striking difference between structure's behaviour is observed with respect to both: time scale and deformation character. More, it is seen that for load above critical deformation practically does not depend on the magnitude of imperfection, which is not a case for loads below Euler's force.

If damage growth is taken into account, then there is no significant difference in displacement pattern as for both cases damage can be neglected, but for different reasons. For load below Euler's force the damage development is very slow (cf. Figure 8), and comes to essential growth just when displacement suddenly rise almost instantaneously. For loads above Euler's force, all process occurs in so short time (cf. Figure 9) that damage reaches only low values during whole process time-span (less then 0.2).

5 TWO-BAR STRUCTURE

The effect of coupling between displacement and damage is much more pronounced in the case of two-bar structure shown in Figure 1b., for which critical elastic load is now 2.88 MN.

121

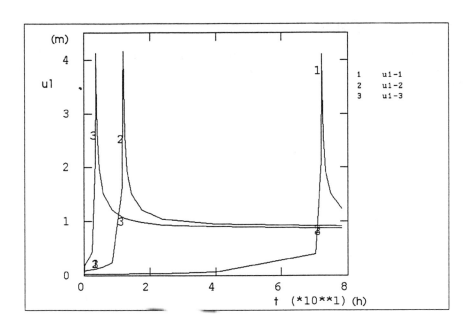

Figure 6. Time dependent lateral displacements without damage for loading P=1 MN

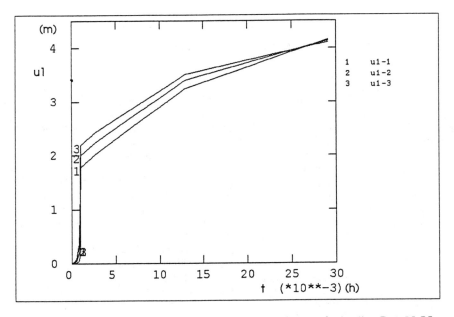

Figure 7. Time dependent lateral displacements without damage for loading P=1.5 MN

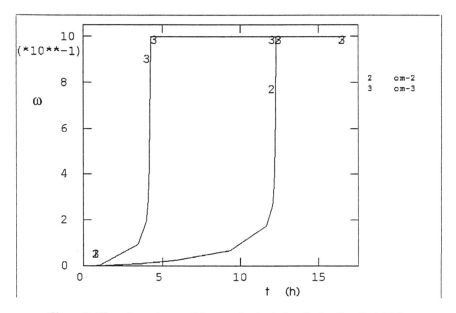

Figure 8. Time dependence of damage in single bar for loading P=1 MN

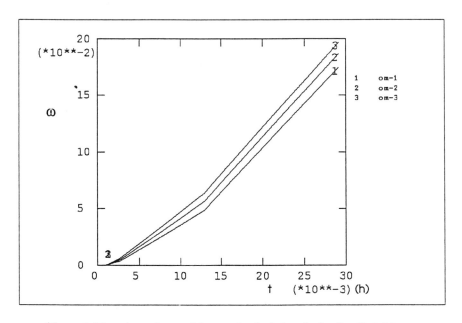

Figure 9. Time dependence of damage in single bar for loading P=1.5 MN

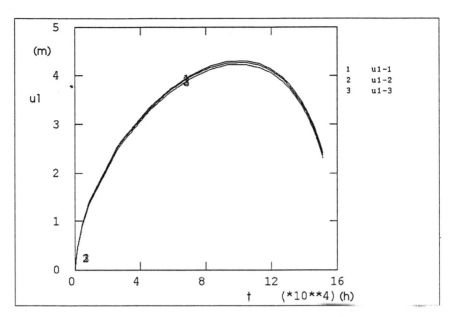

Figure 10. Displacements for two-bar truss for P= 2 MN (no damage)

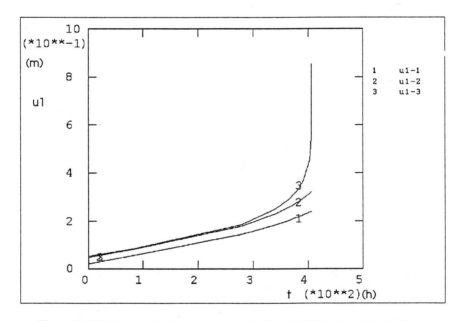

Figure 11. Displacements for two-bar truss for P= 2 MN (damage included)

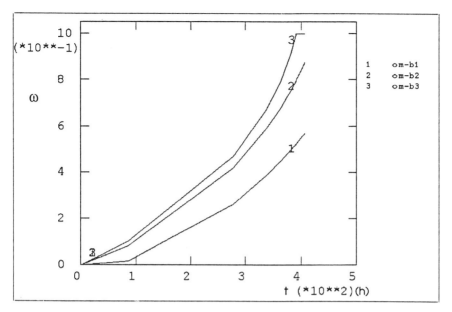

Figure 12. Time dependence of damage in two-bar truss for loading P=2 MN

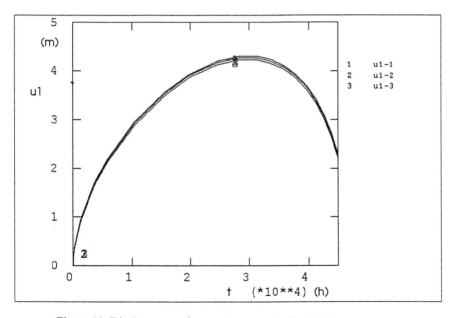

Figure 13. Displacements for two-bar truss for P= 3 MN (no damage)

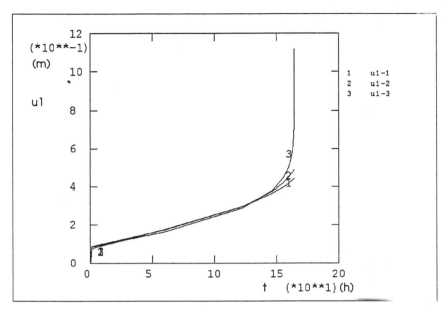

Figure 14. Displacements for two-bar truss for P= 3 MN (damage included)

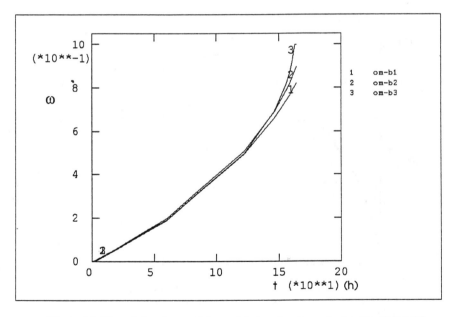

Figure 15. Time dependence of damage in two-bar truss for loading P=3 MN

126

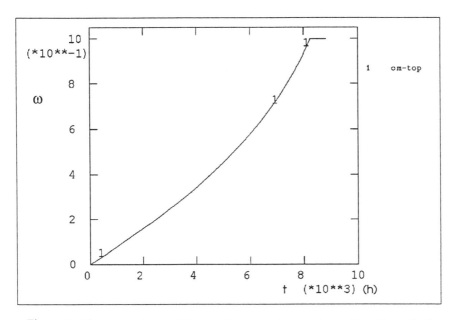

Figure 16. Time dependence of damage for a three-bar structure without imperfections

For load below critical one displacements are shown in Figure 10 (without damage effect) and Figure 11 (with damage taken into account), and damage development in mid-span of buckled bar - in Figure 12. Time scale effect imposed by damage growth is of order $2*10^2$ ($10^5/5*10^2$ hours). Influence of imperfection is seen by comparison of curves labelled 1,2,3 in Figures 11 and 12.

For load above critical one displacements are shown in Figure 13 (without damage effect) and Figure 14 (with damage taken into account), and damage development - in Figure 15. Time scale effect imposed by damage growth is of almost of the same order as in the case of below-critical load: $2.5*10^2$ ($4*10^4/1.6*10^2$ hours). But now influence of imperfection size is negligible (curves labelled 1,2,3 do not differ very much).

6 THREE-BAR TRUSS

For 3-bar structure shown in Figure 4c will be used here to study the influence of general modelling of this structure. In the first example a structure without any imperfection will be studied, whereas in the second case - all three bars will have the same imperfection (initial curvature determined by $e/l=0.004$); damage will be included in both cases.

In the case of structure without imperfection ($e/l=0$) damage occurs only in the upper bar under tension, and its time development is shown in Figure 16. After this bar breaks the structure remains supported by two compressed bars which do not deteriorate (unlimited life-time is expected).

In the case of structure with imperfections in all bars first the upper bar deteriorates at its concave side (curve labelled 1 in Figure 17). Then, as lower bars brake at their convex sides (curve labelled 3 in Figure 17) the damage growth also in convex side of upper bar (curve labelled 2 in Figure 17). All structure eventually fails at time of about $3*10^3$ hours. This time is almost three times less then time in which bar under tension in truss without imperfections has failed ($8*10^3/3*10^3$ hours). This demonstrate importance of taking into account existence of real structures' imperfections.

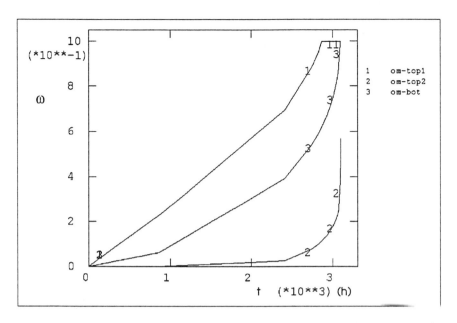

Figure 17. Time dependence of damage for a three-bar structure with imperfections

7 CONCLUSION

The analysis performed has demonstrated that both effects i.e. material deterioration and buckling of structure's members has to be taken into account in the evaluation of lifetime of a structure. It is especially important when imperfections are present.

ACKNOWLEDGEMENTS

The work reported in this paper has been performed under project No. 0378/P4/94/06 granted by KBN (Polish Committee for Scientific Research).
Mr. K. Nowak is acknowledged for assistance in programming under ABAQUS system.

REFERENCES

Ashby, M.F. & Jones, D.R.H. 1986. *Engineering Materials*. Oxford: Pergamon Press.
Bodnar, A., Chrzanowski M., Latus P. & Madej J. 1994. Safety of materials and structures in creep conditions. *Int. J. Pres. Ves. & Piping* 59:161-174.
Bodnar, A., Chrzanowski, M. & Latus, P. 1992. Life-time evaluation of creeping structures. In *Proc. 5th Int. Conf. Creep of Materials, Orlando, 17-21 May 1992:* 461-470
Le May I. 1995. Damage assessment and life extension in aging power plants. *Proc. Central European & World Connection Electric Power Industry, Forum '95, Kraków, 12 -14 October, 1995:* 83-90
Kachanov, L.M. 1958. On time to failure in creep conditions (in Russian). *Izv.Ak.Nauk SSSR OTN* 8:26-31
Hayhurst D.R. 1972. Creep rupture under multiaxial state of stress. *J.Mech. Phys. of Solids* 20:381-390
Lemaitre, J. 1986. Damage constitutive equations. *CISM Course in Damage Continuum Mechanics*, 13-16 Sept, Udine, Italy
Bodnar, A. & Chrzanowski, M. 1991. A non-unilateral damage in creeping plates. *Proc. IUTAM Symp. Creep in Structures IV, Kraków, 10-14,Sept.1990:* 287-293, Berlin, Springer-V.
Chrzanowski, M. & Latus, P. 1996. Coupled effects of damage and buckling in creeping structures. *Proc. 5th Int. Symp. Creep & Coupled Proc., Białowieża, 28-30 September, 1995:*51-56

Ageing of Materials and Methods for the Assessment of Lifetimes of Engineering Plant, Penny (ed.)
© *1997 Balkema, Rotterdam, ISBN 90 5410 874 6*

Prediction of creep cracks in low alloy steel pipe welds by use of the continuum damage mechanics approach

Jan Storesund
ÅF-IPK AB, Stockholm, Sweden

Peder Andersson, Lars Å. Samuelson & Peter Segle
SAQ KONTROLL AB, Stockholm, Sweden

ABSTRACT: Welds are frequently subjected to creep damage in high temperature plant steam piping. Possible mechanisms for two types of creep cracking, type III and IV cracks, are discussed. The formation of such cracks may be influenced by the individual creep behaviour of parent and weld metals as well as the heat affected zone.

A parametric study of different relationships between parent metal, heat affected zone and weld metal creep properties is performed by finite element simulations. By use of a modified Kachanov-Rabotnov constitutive equation implemented into the ABAQUS code and taking the creep ductility into account the damage and lifetime are predicted for a range of circumferential pipe weld creep properties which may have practical importance. The simulations are compared with plant experience. Combinations of creep properties of the constituents in a weld which gives the longest lifetimes as well as the ones which should be avoided are demonstrated. Possibilities to control the weld creep properties for design purposes are discussed.

1 INTRODUCTION

Most of the creep damage found during in-service inspections of high temperature pressure vessel and piping systems is localised to the weldments. Severe accidents have occured, in particular in power plants utilising seam welded pipes. Extensive investigations carried out in order to characterise the susceptibility of welds to creep damage were for instance based on use of the replica technique, (Sandström 1992).

The residual stresses caused by the welding process have often been assumed to cause the damage frequently noted, and that a PWHT will reduce these residual stresses in weldments, thus leading to a homogenised structure. At the same time, much research on creep properties of weld and HAZ materials has been carried out and the results show that severe mis-matching is often present. Typical material data for weldments have been published in (Wang 1993).

It has been shown by use of finite element analyses, (Stevick 1991), that differences in creep properties between the weld and the parent materials lead to stress concentrations of the order of a factor of two in an X-shaped seam weld predicting a severe reduction in the safe life of the pipe. Similar investigations on circumferential pipe welds have been carried out, (Samuelson 1994) and (Eggeler 1994), which show that the creep process produces stress enhancements due to material mis-matching.

The results of FE calculations of the steady state stress distributions were used to estimate the rupture life of a welded pipe, (Tu 1993). A creep damage measure was evaluated based on the Kachanov-Rabotnov theory. It was pointed out that, since the evaluation was based on the steady state stress distribution, the accuracy of the prediction may be low. Stress redistribution is to be expected during the tertiary creep stage. The procedure was later utilised, (Segle 1996a), to estimate the creep life of seam welded pipes.

Creep damage mechanics has been shown to be a powerful tool in prediction of the evolution of creep cracking and creep rupture, (Hayhurst 1996). The creep damage theory of Kachanov - Rabotnov was programmed, (Moberg 1995), and introduced into the ABAQUS FE code. Results of a parametric study, (Segle 1996b), showed the potential of the method to predict creep damage as a function of the creep properties of the weldment constituents. However, the analyses showed only minor differences in creep life due to small differences in the creep properties selected for the study.

This paper presents a parametric study of mis-matched circumferential pipe weldments where the influence of the creep ductility is studied in particular. The results are compared to in-service plant experience of weldment creep damage and creep cracking.

2 CREEP CRACKS IN LOW ALLOY STEEL HEAT AFFECTED ZONES

Cracks appear in circumferential pipe welds of power plant piping. In ferritic steels four types of cracks are identified in terms of a classification (Shuller 1974). Type I and II cracks are located in the weld metal and are formed during solidification or during the subsequent PWHT. Improved welding and inspection techniques have reduced the occurrence of these types of cracks. Type III and IV cracks are located in the HAZ and are frequently found in service in creep loaded low alloy steel steam line welds. Understanding of such cracks are therefore of great importance to improve design and residual life assessment techniques for steam line components.

Type III cracks occur in the HAZ close to the fusion line to the weld metal. The weld thermal cycle results in grain growth and, for low alloy steels, a significant coarse grained microstructure in this region. Such microstructure is associated with high creep strength and low creep ductility in which intergranular cracking is possible during relaxation of residual stresses by creep in a PWHT or in service. In general, type III cracks develop during the PWHT or relatively early in service life.

A common situation is that microcracks are detected in the coarse grained HAZ in replica inspections. These inspections are typically conducted for the first time after a relatively long time in service. In such cases the microcracks may have arisen early in service and retarded in growth as residual stresses relax. However, such damage may grow later in service as a result of degeneration of the microstructure and cavitation by creep. In practice, it may be difficult to sort out the amount of damage which developed in association with stress relief and service exposure.

Steam line pipe welds are multi-pass welded resulting in refinement effects. Typically, 50-80% of the distance along the fusion line through the wall is refined or partially refined depending on the welding technique (Clark 1985, Parker 1995). Cracking related to residual stress relief is not likely to arise in such microstructures. However, circumferential creep cracks have frequently been discovered close to the fusion line in refined microstructures of butt welds and branch welds (Brett 1994). Such cracking appears not only in welds in components like T-joints or pipe line terminals which may suffer enhanced axial stresses due to component geometry or system stresses but also in bulk pipe line weldments after a long term service. This type of cracking has been denoted type IIIa cracking and has been observed only in 0.5CrMoV base material welded with 2CrMo weld metal.

Type IV cracks are located in the intercritial part between base metal and fine grained HAZ where the microstructure has been partially austenitised by the weld thermal cycle. The orientation is parallel to the weld fusion line. Such cracks have predominantly been discovered in welds where bending forces or the influence of geometry such as in branch welds result in enhanced axial stresses. However, type IV cracks also appear in bulk steam pipe butt welds as well as type IIIa cracks. In fact, both types have been observed in the same weld (Brett 1994).

The occurrence of type IV cracks is a life limiting factor for low alloy steel welds and the mechanisms for the development of such cracks have been investigated extensively. It has been found that the creep strength may be significantly reduced in the intercritical HAZ where type IV cracks appear as well as multiaxiality and constraint effects may play an important role. Laboratory investigations on low alloy steel weldments are associated with a high degree of complexity:

i) Accelerated creep tests for constitutive equations are performed at high stresses which may involve other deformation mechanisms than in service.
ii) The composite of various creep properties in a cross weld results in multiaxiality and stress enhancements which are influenced by the specimen geometry. Although such influence may be analysed in cross weld specimens (Kussmaul 1993, Storesund 1995) use of the results translated to the multiaxial creep behaviour in plant is not yet fully understood for many types of welds.
iii) The different parts of the HAZ may be simulated by heat treatment and tested one by one. Although such simulations may produce grain sizes which can represent different parts of the HAZ the time at temperature to produce sufficient amounts of homogenous material for test pieces is much longer than the time in the welding process. Investigations which clarify the precipitation behaviour in the intercritical HAZ as well as simulated microstructures are scarce in the literature and differences may appear (Laha 1995).

3 CONSTITUTIVE EQUATIONS

A modified Kachanov-Rabotnov constitutive equation which accounts for inhomogeneity in creep damage is used (Liu 1991). Neglecting plasticity and primary creep the total strain rate is

$$\frac{d\varepsilon_{ij}^{tot}}{dt} = \frac{d\varepsilon_{ij}^{el}}{dt} + \frac{d\varepsilon_{ij}^{cr}}{dt}, \tag{1}$$

where

$$\frac{d\varepsilon_{ij}^{el}}{dt} = \frac{1+v}{E}\left[\left(\frac{d\sigma_{ij}}{dt}\right) - \frac{v}{1+v}\left(\frac{d\sigma_{kk}}{dt}\right)\delta_{ij}\right], \tag{2}$$

$$\frac{d\varepsilon_{ij}^{cr}}{dt} = \frac{3}{2}B\sigma_e^{n-1}s_{ij}\left[(1-\rho)+\rho(1-D)^{-n}\right], \tag{3}$$

and

$$\frac{dD}{dt} = g\frac{A}{\phi+1}\frac{\left[\alpha\sigma_1 +(1-\alpha)\sigma_e\right]^v}{(1-D)^\phi}, \tag{4}$$

$$D_{crit} = 1-(1-g)^{1/(\phi+1)}. \tag{5}$$

In the equations given above ε_{ij}^{tot}, ε_{ij}^{el}, ε_{ij}^{cr}, σ_{ij} and s_{ij} are the total strain, elastic strain, creep strain, stress and stress deviator tensor, respectively. σ_1 and σ_e are the maximum principal stress and von Mises stress, E and v the modulus of elasticity and Poisson's ratio, D and D_{crit} the damage variable and critical damage. The material creep life is assumed to be fully utilised when D/D_{crit} reaches the value one. α is the material constant relating to the multiaxial rupture criterion which ranges from zero to unity, B, n, A and v are the material constants relating to the minimum creep strain rate and rupture behaviour, g, ϕ and ρ the constants accounting for the inhomogeneity of the damage where ρ represents the volumetric ratio of the damaged phase.

4 MATERIAL MODELLING

In components where geometrical and/or material discontinuities are present, continuum damage mechanics simulations are particularly useful in order to understand the creep behaviour of the component. Using this concept, stress redistribution due to the damage evolution can be taken into account and a more profound understanding of how the component behaves when subjected to creep is achieved (Samuelsson 1994, Tu 1994, Samuelsson 1995, Segle 1996a, Hayhurst 1996).

In the present investigation a three-materials model is used to describe a weldment. The material modelling is based on creep tests of weldments carried out at the Swedish Institute for Metals Research (Wu 1992). The testing was carried out on 1Cr0.5Mo butt welded pipe material at 550 °C. Results from creep tests on parent metal (PM), weld metal (WM) and cross weld specimens with deformations concentrated to and measured at fine grained and refined HAZ microstructures gave the required creep curves for determination of the constitutive equations.

In order to study the influence of differences in creep properties between the weldment constituents the coefficients B and A in equations (3) and (4) are varied (Segle 1996a). Basically, three different combinations, *matched*, *creep-soft* and *creep-hard* weldments are studied. In the matched weldment the minimum creep strain rate of the weld metal equals that of the parent metal. In the creep-soft and in the creep-hard weldments the minimum creep strain rate of the weld metal is higher and lower than that of the parent metal, respectively. Simulations of these three combinations without taking the effect of creep ductility into account have been performed previously (Segle 1996b).

In the present study variation of the ductility is introduced. Simulations are performed with the values 5 (characterised as low), 15 (medium) and 25% (high) elongation at failure, ε_f. Figure 1 shows creep curves for the constituents used in the simulations.

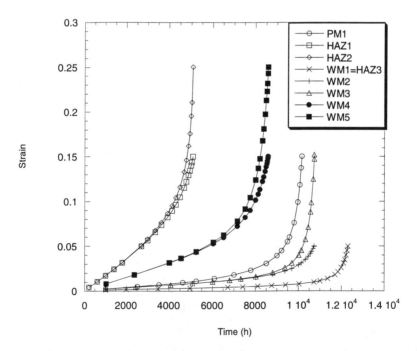

Figure 1. Creep curves used for the simulations based on uniaxial creep tests on 1Cr0.5Mo at 550°C and 110MPa.

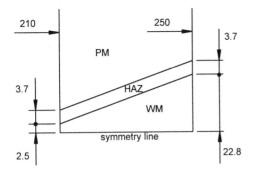

Figure 2. Geometry of pipe with weldment.

Table 1. Minimum creep strain rate, ductility and uniaxial rupture time at 110 MPa for weldment constituents.

Material	Minimum strain rate			Ductility			Rupture time (h)
	Low	Medium	High	Low	Medium	High	
PM1		x			x		10143
HAZ1			x		x		5072
HAZ2			x			x	5072
HAZ3	x			x			12300
WM1	x			x			12300
WM2		x		x			10722
WM3		x			x		10722
WM4			x		x		8582
WM5			x			x	8582

Table 2. Constants in constitutive equations for weldment constituents.

Constant	PM	HAZ1/HAZ2	HAZ3/WM1	WM2/WM3	WM4/WM5
B	1.940e-15	1.540e-13	5.907e-15	1.772e-14	8.860e-14
n	4.354	3.925	3.870	3.870	3.870
A	8.325e-13	4.365e-12	3.304e-13	3.800e-13	4.749e-13
υ	3.955	3.750	4.110	4.110	4.110
g	0.9755	0.9180/0.9848	0.9680/0.9680	0.9220/0.9720	0.8930/0.9385
ϕ	1.423	2.017	0.6517	0.6517	0.6517
ρ	0.393	0.280	0.0985	0.0985	0.0985
α	0.43	0.43	0.43	0.43	0.43
E	160000	160000	160000	160000	160000
ν	0.3	0.3	0.3	0.30.	0.3
ε_f	0.15	0.15/0.25	0.05/0.05	0.05/15	0.15/0.25

5 FINITE ELEMENT MODELLING AND SIMULATION

In the present investigation, the creep damage evolution in a circumferential V-shaped weldment in a piping system is investigated, see Figure 2. The outer diameter and the wall thickness of the pipe are 500 and 40 mm, respectively. The welded pipe is subjected to an internal pressure resulting in a nominal hoop stress of 110 MPa and an axial stress of 50 MPa. Table 1 shows the combinations of minimum creep strain rate and ductility for each weldment constituent which are used in the simulations. The simulations include 11 cases of combinations of the constituents in the weldment.

Table 2 shows the material parameters used for the constituents. It has been suggested, as a result of biaxial testing, that the constant α equals 0.43 for the ferritic steels 0.5Cr0.5Mo0.25V and 2.25Cr1Mo (Browne 1982) why the same value is used for the parent material in the present paper. For the weld and HAZ metals, no data for α were available and hence the same value as that of the parent material is used.

The structure is modelled by use of the 20-node solid element C3D20R and axisymmetrical boundary conditions are utilised.

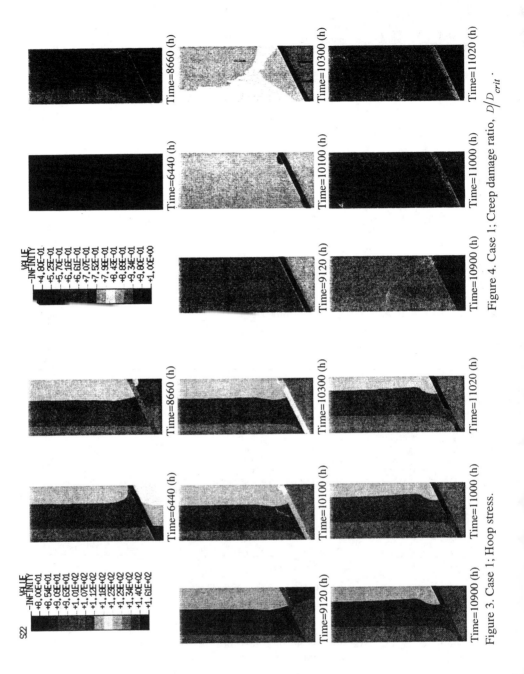

Time=8660 (h)   Time=6440 (h)

Time=10300 (h)   Time=10100 (h)   Time=9120 (h)

Time=11020 (h)   Time=11000 (h)   Time=10900 (h)

Figure 4. Case 1; Creep damage ratio, D/D_{crit}.

VALUE
-INFINITY
+4.80E-01
+5.25E-01
+5.70E-01
+6.15E-01
+6.61E-01
+7.07E-01
+7.52E-01
+7.98E-01
+8.43E-01
+8.88E-01
+9.80E-01
+1.00E+00

Time=8660 (h)   Time=6440 (h)

Time=10300 (h)   Time=10100 (h)   Time=9120 (h)

Time=11020 (h)   Time=11000 (h)   Time=10900 (h)

Figure 3. Case 1; Hoop stress.

S22
VALUE
-INFINITY
+8.00E+01
+8.54E+01
+9.00E+01
+9.63E+01
+1.01E+02
+1.07E+02
+1.12E+02
+1.18E+02
+1.23E+02
+1.29E+02
+1.34E+02
+1.40E+02
+1.61E+02

134

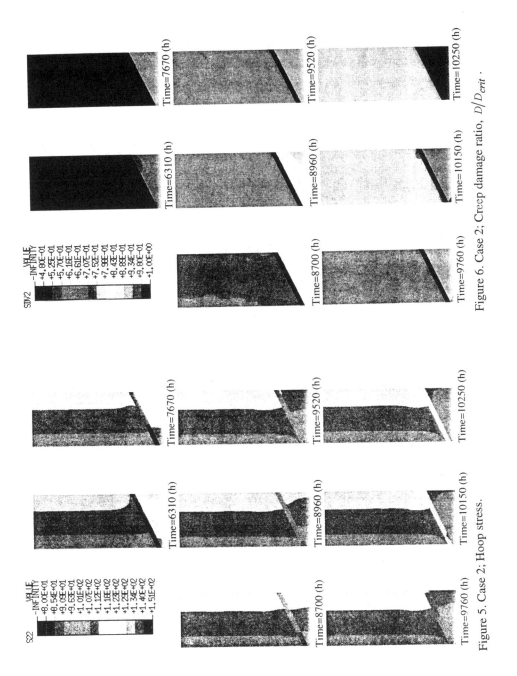

Figure 6. Case 2; Creep damage ratio, D/D_{crit}.

Figure 5. Case 2; Hoop stress.

135

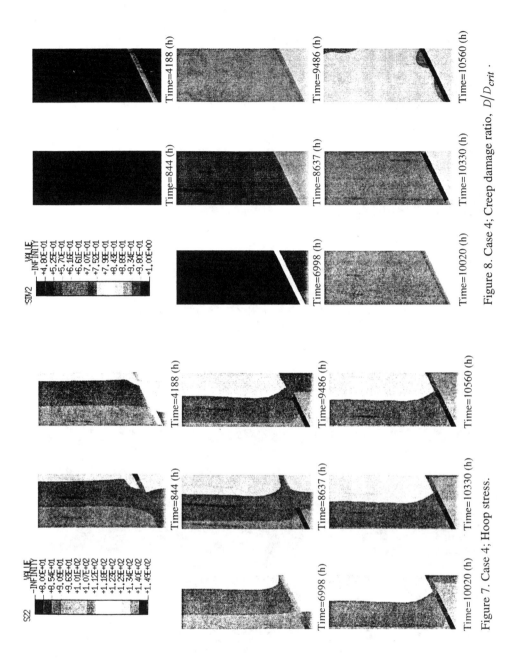

Figure 8. Case 4; Creep damage ratio, D/D_{crit}.

Figure 7. Case 4; Hoop stress.

136

6 RESULTS FROM DAMAGE AND STRESS SIMULATIONS

6.1 Rupture time and positions for simulated cases

The simulations include 11 cases of combinations of the weldment constituents in Table 1. Combination of constituents, rupture position and rupture time for each case is shown in Table 3.

Table 3. Simulated combinations of pipe weld constituents.

Case	Material combination	Rupture position	Rupture time (h)
1	PM1/HAZ2/WM3	Weld metal	11020
2	PM1/HAZ2/WM2	Weld metal	10250
3	PM1/HAZ1/WM2	Weld metal	10220
4	PM1/HAZ3/WM3	HAZ	10560
5	PM1/HAZ2/WM5	Parent metal	10370
6	PM1/HAZ2/WM4	Parent metal	10420
7	PM1/HAZ1/WM5	Parent metal	10370
8	PM1/HAZ1/WM4	Parent metal	10400
9	PM1/HAZ3/WM5	HAZ	9563
10	PM1/HAZ2/WM1	Weld metal	9911
11	PM1/HAZ1/WM1	Weld metal	9910

6.2 Matched weldment

The hoop stress in the matched weldment, case 1, at different times characterising the stress redistribution that takes place in the weldment is shown in Figure 3. Due to a higher minimum strain rate of the HAZ than that of the parent and weld metals, the HAZ is off-loaded. This off-loading begins as soon as the creep process starts.

The first plot in Figure 3 at 6440 hours, shows the hoop stress just before the tertiary creep starts to influence the stresses in the weldment region. The following stress plots show the influence of tertiary creep on hoop stress to the failure at 11020 h. The weld metal is off-loaded and the HAZ is on-loaded. The stress level in the HAZ is significantly enhanced during the final 1000 hours.

In Figure 4 the creep damage evolution is shown. As seen from the damage plots the creep damage is least developed in the HAZ even though the creep rupture strength of the HAZ is lower than that of the two other constituents. The off-loading of the HAZ in the steady state results in a consumed life fraction less than 0.7 at failure although the nominal HAZ rupture time is only 5072 hours and significant on-loading occurs during the final 1000 hours. The damage evolution in the parent metal is somewhat similar to that of the weld metal as expected.

Final failure occurs in the weld metal at 11020 hours. The uniaxial creep test rupture time at the nominal stress, 110 MPa, is 10722 hours for the weld metal. The longer creep life in the circumferential weld is a result of the off-loading in the tertiary stage, even though stresses of about 120 MPa appear in the weld metal in the steady state (the first 6440 hours).

The hoop stress in case 2: a matched weldment with low weld metal rupture elongation, 5 instead of 15%, is shown in Figure 5. The stress enhancement in the weld metal is more slowly off-loaded than in the case of higher weld metal ductility. It is also seen that the on-load of the HAZ is less pronounced. The creep damage evolution, shown in Figure 6, is concentrated to the weld metal in which the failure occurs after 10250 hours. The lower weld metal ductility results in an 8% life time reduction.

A further change by reducing the HAZ ductility from 25 to 15% rupture elongation, case 3, did not have any significant influence on the stress distribution and damage evolution.

Case 4 is a simulation with low ductility, low minimum creep strain rate and a relatively long uniaxial creep test rupture time, 12300 hours, in the HAZ. This may represent creep properties of a weld with coarse grained HAZ. Figure 7 shows the hoop stress for case 4. The first plot shows the steady state creep where the HAZ is on-loaded. Effects of tertiary creep starts just after 844 hours with an off-loading of the HAZ. A simultaneous on-loading in the

137

weld metal can be seen in plots 2 and 3. At 6998 hours the stresses in the HAZ are lower than in the surrounding materials and the off-loading proceeds. The weld metal which was on-loaded starts to off-load at this stage of the creep life. The damage development in Figure 8 shows that the rupture position is in the HAZ. The rupture time is 10560 hours which can be compared to the uniaxial HAZ creep strength of 12300 hours. The excessive stresses in the beginning of the creep life contribute to that reduction in lifetime.

6.3 Creep-soft weldment

Case 5 represents a creep-soft weldment with high weld metal ductility. As for the matched weldment, redistribution of stresses takes place due to differences in creep strain rates between the weldment constituents. In this case, the material discontinuity between the HAZ and the weld metal is reduced compared to the former case. The rupture time is 10370 hours and takes place in the parent metal for which the nominal creep strength was similar, 10143 hours. The HAZ is off-loaded up to approximately 90% of the weld life time. The weld metal is off-loaded in a smaller amount up to 75% of the total life. This results in parent metal failure although the nominal rupture times for the HAZ and weld metal individually are 5072 and 8582 hours, respectively.

Case 6, reduction of weld metal rupture elongation, from 25 to 15%, had insignificant effect on stress distribution and damage evolution. The same reduction in the HAZ or in both weld metal and HAZ, cases 7 and 8, also showed stress distribution and damage evolution similar to case 5. The reason is that steady state creep takes place all the way to rupture in the weld metal and the HAZ with these reductions of the creep ductility.

In case 9 a low ductility and relatively high creep strength HAZ material is introduced in the creep-soft weldment in the same way as for the matched weldment, case 4. Also in this case the rupture takes place in the HAZ due to high on-loading and fast damage evolution in the beginning of the creep life. The rupture time is 9563 hours which is shorter than the one for uniaxial low ductility HAZ that is 12300 hours. The weld metal is significantly off-loaded in the steady state creep. Effects of tertiary creep starts early resulting in a successive on-loading in the weld metal. At the end of the creep life the highest stresses in the weldment are present in the weld metal.

Figure 9 shows the axial stress in case 9. Enhancement of the nominal stress of 50 MPa are present at the outer diameter of the pipe. This enhancement is highest in the HAZ region where stresses up to 94 MPa appear over a long period of the creep life. In the first plot representing steady state creep the high stresses appear in the HAZ and weld metal close to the HAZ. At 4534 hours in the second plot the highest stresses are moved to the type IV region, that is the interface between HAZ and parent metal. The area of this stress enhancement decreases with time and disappears at the end of the creep life.

6.4 Creep-hard weldment

For the creep-hard weldment, the minimum creep strain rate of the weld metal is lower than that of the parent material. An effect of this is that on-loading of the weld metal and off-loading of the HAZ and the parent material in the vicinity of the weld take place when the weldment is subjected to creep. The first plot in Figure 10 shows the redistributed hoop stresses in case 10 at 844 hours which is just before the tertiary stage is entered in the weld metal. The following plots show how the tertiary effects influence the hoop stress field as a function of time. At later stages of the life of the weldment the weld metal is heavily damaged, as seen in Figure 11, resulting in a substantial increase in creep strain rate. This change in strain rate results in the fact that the weld metal is off-loaded, that is at the end of life the creep-hard weldment becomes creep-soft.

The failure of the creep-hard weldment occurs in the weld metal after 9911 hours, see Figure 11, despite the fact that the creep rupture strength of the weld metal is higher than that

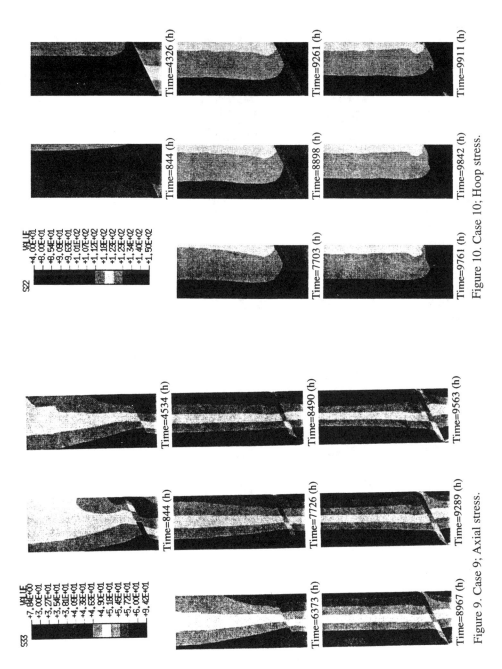

Figure 9. Case 9; Axial stress.

Figure 10. Case 10; Hoop stress.

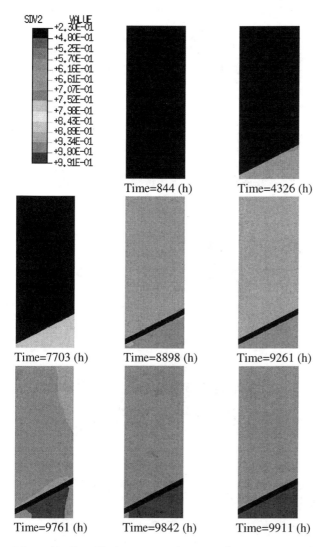

Figure 11. Case 10; Creep damage ratio, D/D_{crit} .

of the parent metal and HAZ. As for the matched and the creep-soft weldment, the explanation for this is the resulting stress redistribution.

The creep-hard weldment shows enhancement in axial stress in the same HAZ regions as for case 9 in Figure 9. The stresses are up to 85 MPA. This enhancement starts in the middle and remains to the end of life.

Simulation with a medium HAZ ductility, case 11, resulted in a stress distribution and a damage evolution similar to case 10.

7 DISCUSSION

7.1 Effects of mis-matched creep properties on hoop stress.

Previous studies have demonstrated that mis-matched weldments in terms of dissimilar parent and weld metal creep strain rates give shorter life than matched ones (Segle 1996b). The differences in life time compared to matched weldments were minor, about 10%, when the strain rates varied with a factor of 3-5. However, the stress distributions and re-distributions as well as damage evolution varied considerably. The present analyses show that low ductility in one of the constituents gives creep life reductions also in the matched weldment, cases 2, 3 and 4, whereas variations in medium and high ductilities have insignificant effects.

The simulations demonstrate that the circumferential weldment is an interacting system where on and off-loadings of the constituents occur due to the current strain rate in parent metal, weld metal and HAZ. Mis-match in minimum creep strain rate results in on-loading of the constituent which display low strain rate. On-loading results in earlier tertiary creep, higher strain rates and, at a later stage, off-loading of the constituent. These interactions are beneficial for the life of the weldment.

It is shown that variation of ductility has a greater effect when this variation also involves variation of the on-set of tertiary creep. In the present model the variation of the on-set of tertiary creep was relatively small. This results in a small influence of ductility in the present cases as long as the ductilities in the weld metal and the HAZ are equal or higher than the one of the parent metal. Simulated cases with small influence of ductility can therefore only represent materials with such creep behaviour.

The mis-match in creep properties chosen in the present exercises is related to what can be assumed to be possible to control in practical welding of similar materials. Larger life reductions can be expected for more extreme mis-matching of strain rates. A practical example of this is repair welds without PWHT which fails prematurely. In such cases the new weld metal can be expected to be considerably mis-matched in relationship to the surrounding, by long term service, degenerated material. The effect may be a severe on-loading which causes failure in a short time.

In addition to weld repairs, further work is needed to analyse e.g. mis-match effects in dissimilar welds where the creep properties of the constituents can differ considerably and premature failures have been reported frequently.

7.2 Mis-match effect on damage evolution

The damage evolution correlate to the on and off-loading effects due to mis-matching. On-loaded weldment constituents may therefore reveal damage such as significant creep cavitation early in creep life. At stages of off-loading such damage may grow slower than expected. Micro-cracks of type III formed by relaxation of residual stresses may also mix up the interpretation of in-service damage in the coarse grained HAZ. On the other hand, constituents which are off-loaded from the beginning of service and then on-loaded may not show creep cavitation until later stages of creep life. In such a case also a rapid damage development could be expected. Stepwise as well as more continuous development of creep cavitation has been observed in service by replica technique (EPRI 1993). It can be suggested that mis-match of the weldment contribute to such stepwise damage development and that understanding of how to make matched welds in practise would improve the accuracy in assessing creep life in welds by the replica technique.

7.3 Effect of axial stress

Axial stress enhancements were present in the type IV region at the outer diameter in a number of the simulated cases. The magnitude of these enhancements, up to 94 MPa, indicates that type IV cracking may be influenced by mis-match of the weldment constituents. With regard to the results, creep-hard welds and creep-soft welds with large amounts of coarse grained HAZ with low ductility, would be more sensitive to type IV cracking than the other cases where the enhanced stresses are about 70-75 MPa. The stress distributions agree with observed growth of type IV cracks in bulk pipe line weldments which becomes significantly slower after an initial fast growth rate in the first 5% of the wall thickness. In the fast growth area also high amounts of creep cavities appear (Brett 1994). The simulations indicate the contribution of stress concentration to the formation of type IV cracking. Proper shear stress creep testing of the material in the type IV zone in real weldments may give answer to the contribution of poor creep properties in that region (Kimmins 1996).

7.4 Effect of stress state

Creep ductility is not only dependent on type of material but also a function of stress state. A high degree of triaxiality in stress can reduce the ductility an order of magnitude or more compared to the uniaxial case (Rice 1969). In regions with material discontinuities, e.g. weldments, constraint effects result in an enhanced stress triaxiality. This is particularly pronounced in HAZ's where a relatively thin layer of material is constrained to follow the adjacent parent and weld material.

The influence of stress state on creep ductility is not considered in the present constitutive equations. In cases where high stress triaxiality reduces creep ductility and thereby the creep behaviour of the weldment, the present model can still be used by choosing a lower value of creep ductility in regions where a high degree of triaxiality is expected, i.e. cases 4, 9, 10 and 11 above.

The simulations do not indicate that type IIIa cracking can be predicted by stress enhancement and damage evolution. One possibility is that the model does not account for reduced ductility due to triaxiality at sharp interfaces as a fusion line when the HAZ and weld metal creep properties are different.

7.5 Prediction of creep cracks

In design and particularly life assessment of high temperature components, the understanding of crack initiation and crack propagation is essential. Where macrocracks emanate from bulk creep cavitation the concept presented in this paper is applicable. With appropriate creep data for the material in the type IV zone and simulations with transverse tensile stresses added to the internal pressure the sensitivity to cracking may be predicted for a range of practical weld performances (Hayhurst 1996). However, for cracks that start from a single defect, for example a fusion line defect in a weldment, the fracture mechanics concept or the continuum damage mechanics concept, using a local approach, must be used.

8 CONCLUSIONS

A parametric study of mis-match in weldment creep properties has been performed by use of a continuum damage mechanics approach. From the results presented in this paper it is possible to draw the following conclusions:

1. The knowledge of the creep properties, involving minimum strain rate, tertiary creep strain rates, rupture time and rupture ductility, of the weldment constituents is essential in order to understand the creep behaviour of a welded component.
2. Stress re-distributions in the weldment are influenced by mis-match of the creep properties resulting in interacting on and off-loadings which can be beneficial for the weldment creep life.
3. Low creep ductility in the weld metal or the HAZ in relationship to the parent metal, results in significant creep life reduction of the weldment.
4. Axial stress enhancement, up to a factor of two, may appear in the same area as type IV cracks typically are initiated in steam line weldments. Mis-match of the creep properties influences the magnitude of these stresses. Some simulated cases, creep-hard weldments and matched ones with low creep strain rate and HAZ ductility display axial stress in a range of

142

77-85% of the hoop stress. At such stress levels type IV cracks may initiate without external axial forces.

5. On and off-loading effects demonstrated in the simulations agree with observations of stepwise creep damage evolution in plant weldments.

REFERENCES

Brett, S.J. 1994. Cracking experience in steam pipework welds in National Power, VGB Conf., Materials and welding technology in power plants 1994 - Essen, March 15&16.

Browne, R.J., Lonsdale, D. & Flewitt, P.E.J. 1982. Multiaxial stress rupture testing and compendium of data for creep resisting steels, *Trans. ASME, J. Eng. Mat. Technol.*, 104: 291-296.

Clark, J.N. 1985. Manual Metal Arc Weld Modelling: Part 1-3. *Mat. Sci. and Technol.* 1: 1069.

Eggeler et. al. 1994. Analysis of creep in a welded 'P91' pressure vessel, Int. J. Pres. Ves. & Piping, 60: 237-257.

EPRI 1993., Life assessment of boiler pressure parts, TR-103377-V5.

Hayhurst, D.R. 1996. High-temperature design and life assessment of structures using continuum damage mechanics, Sixth Int. Conf. on Creep and Fatigue, Design and life assessment at high temperature, London, 15-17 April 1996: 399-410.

Kimmins, S.T., Walker, N.S.& Smith, D.J. 1996. Creep deformation and rupture of low alloy ferritic weldments under shear loading, *Journal of strain analysis*, 31: 125-133.

Kussmaul, K., Maile, K.& Eckert, W. 1993. Influence of stress and specimen size on creep rupture of similar and dissimilar welds, *Constraint effects in fracture. ASTM STP 1171*, E. M. Hackwett, K.-H. Schwalbe, and R. H. Dodds, Eds., American Society for Testing and Materials: 341-360.

Laha, K., Chandravathi, K.S., Rao, K.B.S. & Mannan, S.L. 1995. Creep deformation and rupture behaviour of 2.25Cr-1Mo steel weldments and its constituents (base metal, weld metal and heat affected zones), Proceedings of the 2nd int. Conf. On Heat-Resistant Materials, Gatinburg, Tennessee, 11-14 September.

Liu, Y., Sun, X.F. & Gao, Q, A micro-composite model of localisation in creep damage 1991. Proceedings of Sixth Int. Conf. Material Mechanical Behaviour (ICM-6), Kyoto, Japan (eds. M. Jono and T. Ihone) Pergamon Press, Oxford,. 3: 859-864

Moberg F. 1995. Implementation of constitutive equations for creep damage mechanics into the ABAQUS finite element code, SAQ/R&D-Report No. 95/05.

Parker, J.D. & Stratford, G.C. 1995. Microstructure and performance of 1.25Cr0.5Mo steel weldments, *Materials at High Temperatures*, 13: 37-45.

Rice, J.R. & Tracey, D.M. 1969. On the ductile enlargement of voids in triaxial stress fields, *J. Mech. Phys. Solids*, 17: 201-217.

Samuelson, L.Å., Tu, S.-T. & Storesund, J. 1994. Life reduction in High Temperature Structures Due to Mis-Match of Weld and Parent Material Creep Properties, *Mis-Matching of Welds*, ESIS 17, Edited by K.-H. Schwalbe and M. Kocak, Mechanical Engineering Publications, London: 845-860.

Samuelson, L.Å., Segle, P., Storesund, J. & Wu, R. 1995. Use of the continuum damage mechanics concept in design of piping components working in the creep range, Presented at the 1995 ASME PVP Conference, Honolulu, USA.

Sandström, R., Wu R., & Storesund J. 1992. Through Thickness Distribution of Creep Damage in a Service Exposed Double T-Joint of 0.5Cr0.5Mo0.25V Steel. Pressure Vessel Technology I, Verband der Technischen Überwachungs-Vereine, Essen: 772-789.

Segle, P., Tu, S.-T., Storesund, J. & Samuelson, L.Å. 1996a. Some issues in life assessment of longitudinal seam welds based on creep tests with cross-weld specimens, *Int. J. Pres. Ves. & Piping*, 66: 199-222.

Segle, P., Samuelson, L.Å., Andersson, P. & Moberg, F. 1996b. Implementation of constitutive equations for creep damage mechanics into the abaqus finite element code-some practical cases in high temperature component design and life assessment, Presented at the Symposium on Inelasticity and Damage in Solid Subject to Microstructural Change, September 25-27, 1996 St. John's, Newfoundland, Canada.

Shuller, H. J., Hagn, L. & Woitscheck, A. 1974. Cracks in the weld area formed parts in superheater steam lines-material analysis, *Der Maschinenschaden*, 1: 1-13.

Stevick, G. R. & Finnie, I. 1991. Failure Assessment of Weldments at Elevated Temperatures, Proc. 6th Int. Conf. Mechanical Behaviours, Kyoto, June, 1991.

Storesund, J. & Tu, S. -T. 1995. Geometrical effect on creep in cross weld specimens, Int. J. Pres. Ves.& Piping, 62:179-193.

Tu, S.-T. & Sandström, R. 1994. The evaluation of weldment creep strength reduction factors by experimental and numerical simulations, *Int. J. Pres. Ves. & Piping*, 57: 335-344.

Tu, S. T. & Sandström, R. 1993. Numerical Simulation of Creep Exhaustion of Weldments and Some Design Considerations, Creep and Fracture of Engineering Materials, Eds. B. Wilshire, R. W. Evans, Proc. 5th Int. Conf., Swansea, 27 Mar - 2 Apr,.

Wang, Z. P. & Hayhurst, D. R. 1993. Materials Data for High-Temperature Design of Ferritic Steel Pressure Vessel Weldments, *Int. J. Pres. Ves. & Piping,* 55: 461 - 479.

Wu, R., Storesund, J., Sandström, R. & von Walden, E. 1992. Creep properties of 1Cr0.5Mo steel welded joints with controlled microstructures, *Welding in the world.* 30: 329-336.

Life extension of aged plant

Extending the reliable operation of ageing power stations through analysis of failed components

H.C. Furtado
Centro de Pesquisas de Energia Elétrica, Rio de Janeiro, Brazil

J.A. Collins
Athlone Power Station, Cape Town Electricity, South Africa

I. Le May
Metallurgical Consulting Services, Saskatoon, Sask., Canada

ABSTRACT: Determination of the causes of failure of components in power stations can be very important for avoidance of future problems and unplanned outages. In this paper the authors describe their examination of a number of components that had failed in service, report on the failure mechanisms and indicate the remedial measures or other actions required and taken to ensure continued safe operation and avoid such occurrences in future. Examples include ligament cracking, cracking at austenite-ferrite welds in superheaters and reheaters, cracking from a backing ring, circulating pump cracking, downcomer failure, and turbine casing cracking.

1 INTRODUCTION

When failures occur in power stations the result can be very costly in terms of loss of generating capacity and income for some time while repairs are made, together with the costs of purchasing alternate power and of the repairs themselves. When a station is new there are almost always a number of unplanned shutdowns until the teething problems are overcome, and then the losses tend to stabilize at a low level. The well known bath-tub curve for losses is followed, with an increase in failure rate and lost output occurring as the plant ages and approaches the end of its design life.

In order to reduce the losses that occur as power stations age, it is important to determine the causes of failures and to take remedial action to prevent similar recurrences, either in the same unit or in other stations of similar type. Hence it is important that failure analysis is made following an incident. Not only can this assist in the prevention of other unplanned shutdowns, but the results can assist with redesign to avoid future problems of the same type, if the lessons learned can be transmitted back to the designers and manufacturers of the plant.

In this paper the authors describe a number of investigations of boiler components that failed in service and the lessons to be learned from them.

2 CASES

2.1 Ligament cracking

In 1995, during a life extension exercise undertaken after 120,000h operation of this 25 MW boiler, it was decided to renew the primary superheater hanger tubes from just below the boiler roof down to the inlet header. Radiographs made of the inlet port header stub-to-tube welds taken at this time indicated there were cavities or pores in the header stubs, nowhere near the welds. Selected welds were cut out, confirming the presence of internal corrosion pits. Accordingly, it was decided to replace all the inlet port header stubs.

When the stubs were removed and the attachment areas were being spot faced for attachment of new stubs, cracking on the header was identified and it was scrapped. Figure 1 shows a section of the header with the inlet stubs removed and the end of an outlet port stub (larger diameter) in place. Figure 2 shows the internal ligament cracking (a) as observed directly, (b) as shown up by means of fluorescent magnetic particles and (c) as indicated using liquid penetrant. It may be seen how the fluorescent magnetic particle inspection shows up the very large number of surface cracks, while liquid penetrant shows only the major ones. It may be noted that no cracks were apparent on the outside of the header and that they were detectable only on its inside and in the bores of the stubs.

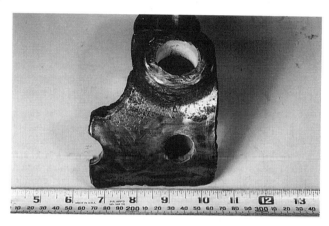

Figure 1. Section of the header with inlet stubs removed and the end of an outlet stub still in place (upper part).

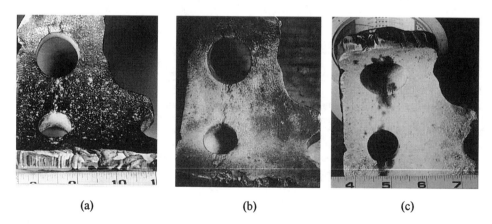

(a)   (b)   (c)

Figure 2. Internal ligament cracking in the header; (a) as observed directly; (b) as shown by fluorescent magnetic particles; and (c) shown with liquid penetrant applied.

Figure 3 shows the cutting sequence followed in preparing metallographic specimens. The cracking is shown in Figure 4, and at higher magnification in Figures 5 and 6.

From the appearance of the multiple branched cracks, filled with oxide and corrosion products, it was concluded that the cracking initiated from thermal fatigue, and the cracks grew from thermal cycling accentuated by stress corrosion cracking (SCC) during shutdown. The cracks initiated in the mild steel header, which operates at approximately 285°C and 4.38MPa. The primary problem is that the boiler had operated with frequent shutdowns as it had been used for

148

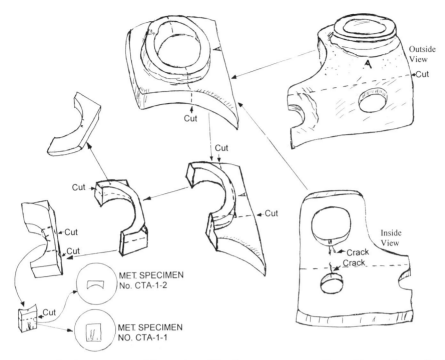

Figure 3. Cutting sequence of the section of header to prepare metallographic specimens for detailed examination.

| (a) | (b) |

Figure 4. Cracking observed in specimens CTA-1-2(a), and CTA-1-1 (b).

peak load requirements, and the design did not consider this. The evidence of SCC in the mild steel header indicates that oxygenated water had been present for a significant period of time. It suggests that the long term storage to which was boiler was subjected (1986-1995) was not as effective in eliminating corrosive conditions as was originally thought. Also, it seems probable that the boiler water treatment used in the early life of the station in the 1960s was not made with adequate knowledge and care. Procedures for preserving steam generating plant during shutdown and avoiding corrosion problems have been discussed previously (Collins 1996).

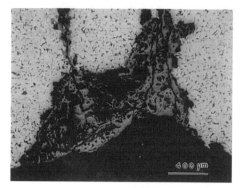

Figure 5. Oxide filling the mouth of a crack.

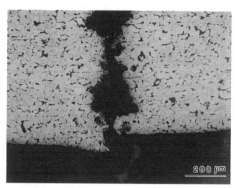

Figure 6. Displacement caused by cyclic strain at the welds surface.

2.2 *Austenite-ferrite weld cracks*

An evaluation was made of the boiler of a 220MW generating unit that had operated for some 30,000h over a 23 year period, with frequent shutdowns and much of its operation on part load. Sample welds between the stubs of ASME SA213 T22, 1¼Cr, 1Mo steel and the austenitic tubes of the secondary superheater and reheater were removed for examination, although nothing amiss had been noted during nondestructive examination. The secondary superheater tubes were of AISI 347H and the reheater tubes of AISI 304H. The samples examined included shop welds between the stubs and austenitic steel nipples and field welds between these and the tubes.

Figure 7 shows a junction between a ferritic stub and a nipple of 347H: there was cracking along the fusion zone on the stub side (Figure 7a), while surface cracks were observed on the stub itself adjacent to the fusion zone crack (Figure 7b). Ahead of these cracks there was cavitation. The cracking was indicative of thermal fatigue promoted by differential expansion between the two materials and the cyclic operation. The cavitation ahead of the parallel surface cracks indicates a local creep mechanism ahead of the crack tips: the cracks would grow to line up with sub-surface voids. It was also observed that intergranular cracking along the fusion line was occurring internally, as shown in Figure 8. The cracks grew along carbides lying intergranularly.

Approximately one-third of the wall thickness was cracked. Similar cracking was observed in other samples. At the welds between the austenite nipples and the austenitic tubes, cracking was also observed (Figure 9).

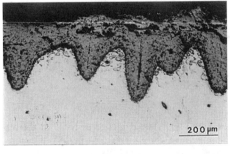

(a)

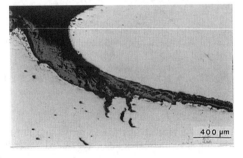

(b)

Figure 7. The junction between a ferritic stub and a 347H nipple. (a) Cracking on the fusion line on the stub side (stub on left). (b) Surface cracks on the stub adjacent to the cracking.

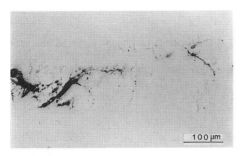

Figure 8. Intergranular cracking along the fusion line occurring internally.

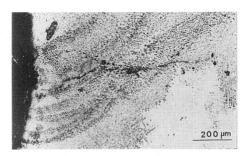

Figure 9. Cracking between the austenitic weld metal and the austenitic nipple.

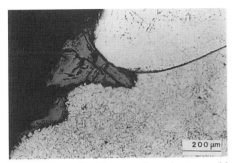

Figure 10. Cracking at the reheater welds between the ferritic stub and the austenitic weld metal.

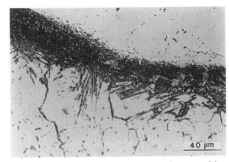

Figure 11. Internal intergranular cracking adjacent to segregated carbides in the reheater.

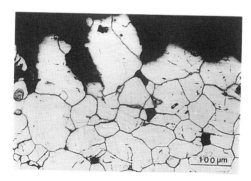

Figure 12. Sensitized microstructure in the 304H reheater tubing.

The reheater connections showed similar cracking between the ferritic nipple and weld metal (Figure 10), with internal intergranular cracking adjacent to segregated carbides (Figure 11). The 304H tubing was completely sensitized with many surface cracks apparent (Figure 12).

It was clear that the cyclic operation had caused serious problems at the ferrite-austenite connections. The extent and seriousness of the cracking disclosed by the sampling indicated a need to replace the connections. The example also demonstrates the danger inherent in operating a boiler designed for continuous operation on a cyclic, peak-load basis. A final point to be made is the difficulty in detecting cracking at ferrite-austenite connections using standard NDE procedures: the removal of samples for destructive examination at appropriate intervals is the only secure method for evaluation.

151

2.3 *Cracking at a backing ring*

This failure was from the same power plant discussed in Section 2.1. Failure occurred at a weld in a 356mm outside diameter steam range. This is a common continuous steam transfer pipe fed from eight boilers and from which six steam turbines take their supply. When installed in the early 1960s it was common practice to use backing rings, and the rupture occurred there. The steam range had operating conditions of 490°C and 4.28MPa. Figure 13 shows a section through the failed weld. In addition to the fracture, another crack has originated at the root of the weld behind the backing ring.

Closer examination showed the cracks to be of SCC type, although they may have been aided by thermal cycling. Figure 14 shows the small crack at the weld root, while Figure 15 shows the branched, intergranular attack, at secondary cracking on the fracture face.

Figure 13. Section through the failed backing ring weld.

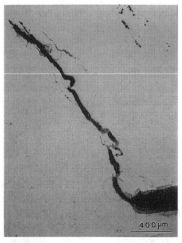

Figure 14. Small crack at the weld root shown in Figure 13.

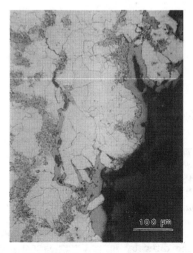

Figure 15. Branched intergranular cracking on the fracture face.

152

Apart from the high stress concentration present at the weld root because of the backing ring, the evidence of SCC was again indicative of problems during shutdown when the system was not dried out completely. Also, a notch such as is produced by a backing ring provides an ideal site for ions to concentrate and SCC to initiate.

2.4 Boiler circulating pump cracking

In the boiler discussed in Section 2.2, one of the circulating pumps was checked and found to have extensive internal cracking on the casing. The "craze" appearance and the depth of 6 to 8mm indicated that this was due to rapid thermal fluctuations, and thermal fatigue enhanced by corrosion effects.

The problem was attributed to excessive leakage of the colder seal water into the boiler water, with stratification providing thermal cycles. The extensive nature of the cyclic operations is thought likely to have contributed to unbalanced loads and excessive wear on the bearing and seals of the pump, giving rise to the condition.

Once more, it appears that equipment designed for steady operation wears or fails prematurely when in intermittent service.

2.5 Downcomer failure

Failure occurred in the downcomer of a boiler that had operated for around 89,000h over a 28 year period, with 722 hot starts and 115 cold starts. The material was DIN ST 45.8, and the operating conditions in the downcomer were 277°C and 5.9MPa. The region where the rupture took place had been destroyed, and examination was made of distorted tube parts that came from areas close to the failure, as well as of adjacent tubes.

The microstructure was spheroidized pearlite and ferrite and the mechanical properties were appropriate for the specification. However, heavy external corrosion was observed, together with internal pitting. The nominal tube wall thickness was 4.5mm and areas near to the rupture had thicknesses as low as 1.8mm. Adjacent downcomers were also heavily corroded. Accordingly, it was concluded that the problem was one of external corrosion and poor water treatment leading to internal attack, failure being because of reduced wall thickness.

Replacement of the downcomer was required and avoidance of future problems depends on good control of the boiler water together with monitoring of external corrosion. The problem would have been avoided if periodic thickness measurements had been made.

2.6 Turbine casing cracking

During a major inspection of a 300MW turbine that had operated for 17 years with an inlet steam temperature of 538°C, cracking was observed in a groove adjacent to the steam inlet; this was a region where the temperature would have been close to the maximum. The majority of the cracking was confined to the upper half of the casing. It was reported that no cracking had been observed during the last inspection, six years before.

When the upper half of the casing was examined after NDE inspection using liquid penetrant had disclosed the cracking problem, it was indicated that the cracks had been ground out. However, examination by the authors using the fluorescent magnetic particle technique showed that cracking was still present. When the cracking was finally removed, the ground out region extended virtually round the 180° of the upper half, and had a maximum depth of 51mm in the 75mm thick wall of the casting.

The causes of cracking in the HP and IP section casings of steam turbines are given (Rasmussen 1986) as:

- Low-cycle/thermal fatigue, 65%
- Brittle fracture, 30%
- Creep, 5%

Thermal fatigue cracks grow slowly in a transgranular manner until a critical size is reached, when fast, catastrophic failure can result. They are normally detected and dealt with during periodic inspections every five years or so, when they occur. Brittle fracture cracks grow much faster, again by a low cycle (thermal cycle) fatigue mechanism, and may be intergranular or transgranular. They occur when the toughness of the casting has been degraded by temper embrittlement during service. Creep damage in castings is generally observed to have grain boundary cavities or voids apparent when polished and etched, cracks developing in an intergranular manner.

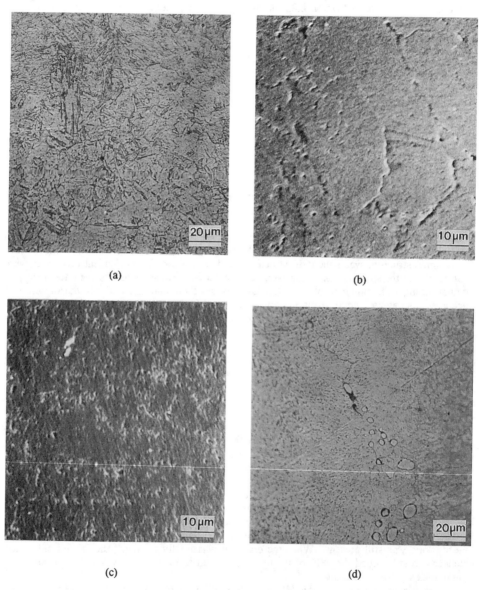

(a)

(b)

(c)

(d)

Figure 16. Typical microstructures from replicas of the turbine casing. (a) Light micrograph showing acicular structure. (b) SEM view showing grain boundaries outlined by carbides. (c) SEM view showing widely distributed carbides. (d) Light microscope view showing area of large carbides with some decohesion apparent at their boundaries.

154

In the present case, the appearance suggested low-cycle thermal fatigue. The crack growth appeared to have been rapid according to the information supplied, namely that no cracking had been present six years previously, but two factors cast doubts on this observation. First, there was evidence of prior grinding in the bottom of the groove that had been oxidized subsequently. It may have been done six years previously or earlier, presumably to remove surface cracks. Second, based on the fact that fine (deep) cracks were missed using liquid penetrant, such cracks could have been missed previously. In this connection, the NDE techniques listed for steam turbine components do not include the less sensitive liquid penetrant procedure according to other sources (Mayer and König, 1988).

Assessment of the state of the material was made from hardness measurements and metallographic replicas, a number of areas being examined adjacent to and distant from the cracks. Because of their removal, it was not possible to replicate the cracked areas themselves.

The material was a modified version of ASTM A356, Gr.8, with higher C (0.10-0.20%) and Ni ($\leq$ 0.50%) content. The as-supplied hardness was specified as 170-220HB. While no specification was given for microstructure, it would be expected to be ferrite and pearlite (probably with partial spheroidization of the latter during tempering) (Viswanathan, 1989).

In all areas, except one, the hardness exceeded the original specifications. The area of lowest hardness (167HB average) was closest to the steam inlet. The microstructures in all cases were of acicular ferrite with some variations in the carbide distribution. Figure 16 shows typical microstructures. No evidence of microstructural problems or of creep voids was detected and only in one isolated area were large carbides with some possible decohesion observed (Figure 16 d).

It was concluded that the cracking had developed from thermal fatigue as the unit had been used for peak load in recent years, with frequent shutdowns. The material was considered sound, and weld repairs, with appropriate heat treatment, were undertaken to put it back into service.

3 CONCLUDING REMARKS

The components examined and reported on here have failed, in many cases, because of a change in operating mode from continuous to cyclic. This applies to the ligament cracking, the austenite-ferrite weld cracks, the circulating pump and the turbine casing. The backing ring failure may be attributed to this cause in part. Improper water treatment and corrosion control played the major role in the downcomer failure and a part in the ligament cracking and backing ring failure.

The cases observed are helpful in the evaluation of other plants as they provide pointers to specific areas to be checked and also emphasize the limitations on some of the inspection techniques commonly employed.

REFERENCES

Collins, J.A. 1996. A practical perspective on plant preservation and plant life extension in thermal power stations, considering aspects of reliability, maintainability, cost and safety. *Proc. CAPE '95 Colloquium on ageing and plant life assessment, Int. J. Pres. Ves. & Piping* 66:27-32.

Mayer, K.H. and König, H. 1988. Determination of residual life of steam turbine components. *Proc. Int. Conf. Life Assessment and Extension*, Section 1: 211-218. The Hague: CRIEPI/EPRI/KEMA/VGB.

Rasmussen, D.M. 1986. Steam turbine case repairs to extend operating life. In R.Viswanathan (ed.), *Workshop proceedings: life assessment and repair of steam turbine casings*, Rept. CS 4676 SR; 8.1-9.1. Palo Alto: EPRI.

Viswanathan, R. 1989. *Damage mechanisms and life assessment of high-temperature components*. Metals Park, OH: ASM International.

Ageing of Materials and Methods for the Assessment of Lifetimes of Engineering Plant, Penny (ed.)
© *1997 Balkema, Rotterdam, ISBN 90 5410 874 6*

Lifetime assessment and repair of steam turbine casings and valve chests

K. H. Mayer, H. König, D. Weber & M. Weiss
GEC ALSTHOM Energie GmbH, Nürnberg, Germany

ABSTRACT: The residual life of turbine casings and valve chests is essentially defined on the basis of the virgin conditions and the actual (but relatively unknown) life history. These complex tasks can only be solved by extensive knowledge of manufacturing technology and the typical appearance of manufacturing flaws, by detailed information on the stress levels of individual zones and the resulting types of damage and by the know-how acquired over many years of experience in practical analysis techniques, risk evaluation and refined repair methods.

The life potential imparted in the virgin state can be viewed in terms of the contractual theoretical service life based on the state of the art at the time of design, and in terms of the manufacturing techniques available at the time. This life potential must be compared with the service history which is generally defined by the actual service conditions and operational faults and cases of damage, etc. Due to the lack of information and the involved nature of the service life analysis, it is clear that theoretical investigations and probability forecasts alone will not yield accurate results without taking stock of the actual conditions of the essential components. This is best accomplished by subjecting them to non-destructive and destructive testing.

The inspection of long-term-stressed steam turbines of fossil-fired power stations frequently brings forth cracks in the casings and valve chests of the HP and IP turbines. The cracks, as a rule, are located in regions of cross setional transitions and at casting or weld defects. They are attributable to high thermal stresses occasioned by transient operating conditions.

Experience has shown that the castings - (unless the cracks can be removed by grinding) - can be reconditioned by weld repair. The question of whether the weld repair can be carried out in situ, or whether it should be carried out in the workshop or by a qualified foundry, must be decided on the merits of each individual case.

INTRODUCTION

Experience with long-term-stressed turbine casings and valve chests of HP and IP turbines shows that the inspection findings are largely linked to the type of operation and to the state of the art at the time of design, manufacture, inspection and commissioning (for example Woodcock 1987).

The steel castings of turbines built in the 1950s were required to handle many more functions than the turbine casings and valve chests of modern turbines. From the point of view of casting and inspection this resulted in complex geometrical shapes which, with regard to the modern-day practice of operating the turbines in the medium and peak load range, did not offer good thermoelastic properties. Where a long-term-stressed casing is subjected to nondestructive inspection during overhaul of the steam turbines according to the present state of the art, it is common practice to find defects in the castings which by far exceed the current limits. As a matter of experience, the cracks which developed during operation are attributable to high thermal stresses and/or casting defects. However, one point of significance is that there has not been a single incident where the cracks have resulted in brittle failure or an outage.

In order to continue satisfactory operation of a casing after repair weld followed by heat

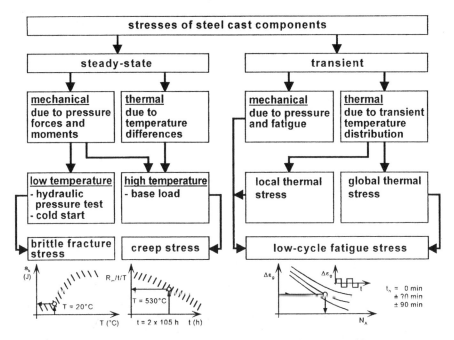

Figure 1. Stresses of cast steel components in steam turbines.

treatment and operational measures, it is necessary that adequate knowledge is available on the factors of influence on further operational safety, such as stress conditions, behaviour of defects, properties of the materials, weldability and dimensional stability.

STRESSING OF TURBINE CASINGS AND VALVE CHESTS

The various types of stresses are, in principle, very similar to those of hot pressure vessels and are shown schematically in Figure 1 (Mayer, 1980).

A distinction must be made between steady-state and transient stresses, caused in each case by mechanical or thermal stresses. A factor of importance for the steady-state stresses is the mechanical stress due to the internal pressure which results in creep stresses at temperatures in excess of 400 °C. Up to now we have not experienced any creep damage in our turbine casings and valve chests. This reflects the relatively high safety factors which are specified for cast turbine components.

In the low-temperature range there are exceptional cases where a brittle failure can occur during hydrostatic testing and cold starting. The transient mechanical stresses are caused by pressure variations and fatigue. Superimposed on these mechanical stresses is a transient/thermal load, resulting in low-cycle fatigue stress.

Highly stressed zones due to thermal operating stresses of box-shaped valve chests of older design are shown as an example in Figure 2.

A particular risk potential for cracks has been found to exist in areas of local concentrations of expansion due to factors associated with the shape of the casting. For instance, additional stresses due to thermal bending at corners, e.g. at "A", or the obstructed expansion and deformation at the transition of a "quasi-rigidly clamped" partition to the casing wall, e.g. at "C", are likely to result in cracks. The same applies to thin stiffening ribs in contact with steam on both sides and which heat up at a faster rate than the environment.

The measure for such stresses as shown in the upper part of the figure for point "A" is the total expansion amplitude $\Delta_{\varepsilon\ total}$ of an operating cycle - due to startup, steady-state operation and rundown - as well as the stress time in the individual phases of the cycle.

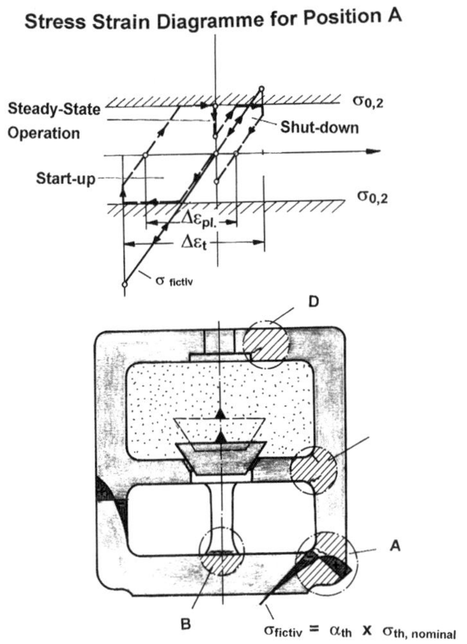

Stress Strain Diagramme for Position A

Steady-State Operation

Shut-down

$\sigma_{0,2}$

Start-up

$\sigma_{0,2}$

$\Delta\varepsilon_{pl.}$

$\Delta\varepsilon_t$

σ fictiv

D

A

B

$\sigma_{fictiv} = \alpha_{th} \times \sigma_{th, nominal}$

Figure 2. Thermal fatigue stresses of a valve chest (schematic).

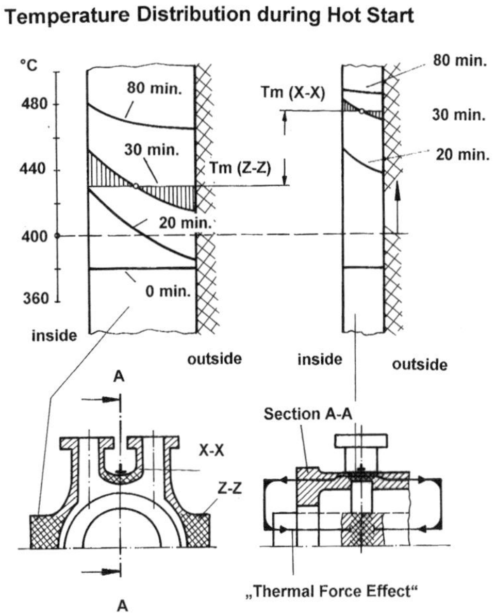

Temperature Distribution during Hot Start

°C

480

440

400

360

80 min.

30 min.

20 min.

0 min.

T_m (Z-Z)

T_m (X-X)

80 min.

30 min.

20 min.

inside

outside

inside

outside

A

X-X

Z-Z

A

Section A-A

„Thermal Force Effect"

Figure 3. Thermal fatigue stresses of a turbine casing due to global temperature distribution during a hot start.

159

In addition to these locally limited thermal stresses, associated directly with the temperature difference in the area under review, there are also cases where the temperature differences developing over large areas of the component under transient operating conditions, e.g. in a turbine casing (Figure 3), cause additional thermal stresses between the flanges Z - Z and the web between the inlet connections at X - X.

These examples make it clear that the designer of steam turbines for cyclical operation cannot succumb to giving the casings and valve chests any shape he would wish; he must make sure that the components offer good thermal flexibility and low stress concentration. By doing so we also obtain more calculable stress patterns, cast components with fewer casting defects and better interpretation of nondestructive test results.

TYPICAL CASTING DEFECTS

For an appraisal of the scope of inspection and repair of stressed cast steel casings, specific knowledge is required of the development and location of the major casting defects. Figure 4, an IP turbine casing half serving as an example, shows the zones and regions of the casting which have a tendency to develop shrinkage cavities, hot tears and surface flaws as well as modern foundry practice which contributes towards prevention of such defects by appropriately positioned and dimensioned feeder, padding and external chills (Mayer 1980).

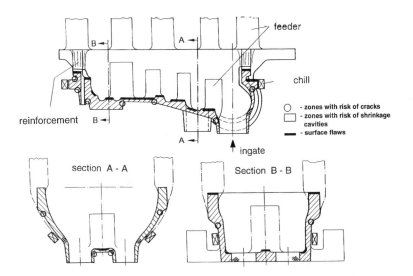

Figure 4. Casting areas with increased defect risk.

- Shrinkage cavities develop due to shrinkage, i.e. due to the volume deficit between the liquid and solid state. They can largely be avoided by correct feeding, i.e. by so-called controlled solidification. Constant wall thicknesses, local material accumulation and so-called hot spots are at risk.
- Hot tears develop while the casting is cooling down in the temperature range between liquid and solid, i.e. while it is in a doughy state just above the temperature at which the material solidifies completely. The cracks develop on account of obstructed shrinkage at cross-sectional transitions, at so-called hot spots (spots which solidify at a slower rate, e.g. fillets and at external chills). The cracks are particularly dangerous because they are perpendicular to the surface and start just beneath the surface; we have found that the hot tears are the main cause of operational cracks.

In Germany a joint programme is under way in which fracture mechanics investigations are

160

being made on larger specimens of hot-torn test castings of the temperature-resistant 1% CrMoV cast steel GS-17 CrMoV 5 11 (Barens 1994) - Figure 5. The results hitherto available reveal good agreement with the findings established for the specimens of the flaw-bearing forging, both in respect of the fracture mechanics behaviour and also with regard to agreement between the flaw size predicted on the basis of the echodynamic results and the actual flaw size.

- Surface flaws are defects in the form of accumulations of non-metallic inclusions. They occur predominantly in areas where the rising hot metal contacts the underside of the moulds and cores.

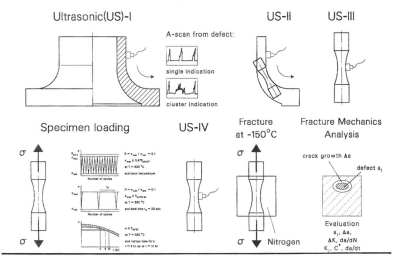

Figure 5. Schematic presentation of selection, stressing and evaluation of specimens of defect-bearing castings.

NONDESTRUCTIVE TESTING OF HEAT-RESISTANT CAST COMPONENTS

An inventory of the condition of long-term-stressed casings of cast steel must take into account the test standards which prevailed at the time the castings were made. In this connection it is of interest to make a historic review of the relevant DIN standards, even though higher testing standards were at times specified by the turbine manufacturers.

The most important details are: During the period from 1942 to 1959 a surface crack detection test according to DIN 1682 was carried out on request. This DIN standard did not include a volumetric nondestructive test of the castings. The first edition of DIN 17 245 "Heat-resistant ferritic steel castings" gave no details whatsoever on nondestructive testing.

The second edition of DIN 17 245 issued in 1964 mentioned the magnetic particle test as being suitable for examining castings for "surface defects". This edition also contained details on the maximum permissible size and frequency of defects along the lines of reference radiographs in ASTM-E 71. Ultrasonic inspection was adopted to supplement radiographic inspection and, from 1967 onwards, ultrasonic inspection was allowed to be used instead of radiographic inspection. Since 1977 comprehensive testing instructions have been given in DIN 17 245. A Stahl-Eisen Test Code (SEP 1922) was issued for ultrasonic testing; reference to this is made in DIN 17 245.

Present-day nondestructive testing of turbine casings and valve chests calls for a 100% magnetic particle test and a 100% ultrasonic test. The radiographic test is only performed to supplement the ultrasonic test if the findings cannot be satisfactorily interpreted. The ultrasonic test provides relatively dependable information on non-healed and partly healed thermal cracks (Figure 6) (Christianus 1987).

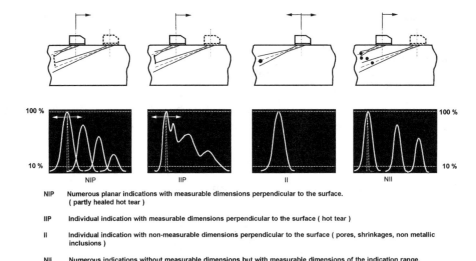

NIP	Numerous planar indications with measurable dimensions perpendicular to the surface. (partly healed hot tear)
IIP	Individual indication with measurable dimensions perpendicular to the surface (hot tear)
II	Individual indication with non-measurable dimensions perpendicular to the surface (pores, shrinkages, non metallic inclusions)
NII	Numerous indications without measurable dimensions but with measurable dimensions of the indication range. (pores, shrinkages, non metallic inclusions)

Figure 6. Typical echodynamics of sub-surface indications in steel castings.

The standard inspection of highly stressed cast components of steam turbines during overhaul constitutes a visible inspection of the whole casting and a magnetic particle test of a high stressed area. Ultrasonic tests are only performed to supplement the magnetic particle test to determine the depths of the cracks etc.

MATERIAL PROPERTIES AFTER A SUSTAINED PERIOD OF OPERATION

Before deciding whether the cracked casings can be used for any further period of time, it must be examined whether repair welding will ensure continued suitability for operation. For this purpose, test specimens are taken from the vicinity of the repair zone for:

- tensile testing
- notched bar impact energy testing (FATT)
- low-cycle fatigue testing, and
- creep testing

As a rule standard test specimens of normal size were used so as to achieve compatibility of test results with the new condition.

Figure 7 shows an example of the transition temperatures (FATT) obtained on impact energy test specimens for:

- operation-stressed condition, and
- post-weld heat treatment condition, i.e. for the repair-welded state.

The maximum transition temperature for the operationally stressed condition lies at about +130 °C. This temperature is reduced on average by about 10 °C - 20 °C as a result of post-weld heat treatment. Based on these results, it is recommended that the preheating temperature should be at least +200 °C both for repair welding and for startup operations.

Figure 8 illustrates the resistance to low-cycle fatigue on samples of 9 casings tested at 20 °C and 530 °C. All the results still agree with the scatter bands applicable to the virgin condition, in fact both to the long-term operationally stressed and the heat-treated conditions. This means that for the repair-welded condition the strength applicable to the virgin condition, or at least the lower limit of the scatter band, is restored. In principle, similar results were also obtained in the case of creep tests.

162

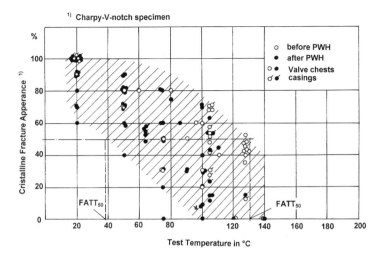

Figure 7. FATT$_{50}$ of long-time service stressed valve chests and casings of 1%CrMoV cast steel.

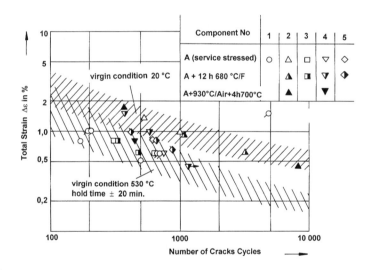

Figure 8. Low-cycle fatigue strength of long-time service valve chests and casings of 1%CrMoV cast steel.

WELD REPAIRS OF STEAM TURBINE COMPONENTS

In deciding whether a cracked casing should be subjected to weld repair, there are numerous questions which must be clarified first:

- Identification of the cause of the crack (casting defect, weld repair defect, thermal stress crack or creep crack)
- Examination of crack pattern by nondestructive testing
- Fracture mechanics behaviour of the crack according to its position, size, depth and direction in the casting
- Conditions of the material in the neighbourhood of the crack (degree of damage, weldability, casting conditions etc.)

163

- The influence of normal and optimized operating conditions on supercritical crack propagation, e.g. improvement by pre-heating during startup and by an optimized starting procedure
- Time and cost of the weld repair versus short-term re-manufacture
- Advantages of improved component design and performance, if provided new.

Based on our own experience over the last 10 to 15 years the following generally valid statements can be made:

- The crack depth and the exact crack pattern in the casting can very often not be predicted with good accuracy by nondestructive testing on account of the shape and condition of the casting and sometimes due to poor access. Consequently, an evaluation by fracture mechanics is very uncertain if the crack is left in place.
- The cracks develop due to local damage on account of high thermal stresses in stress concentration zones. The cracks are frequently encouraged by casting and welding defects.
- More than 50% of all cracks can be removed by grinding with a low-stress-concentration transition radius because the casings are overdimensioned to withstand the stresses of the internal pressure and moments.
- The condition of the material in the neighbourhood of the crack allows a weld repair to be made with observance of the necessary pre-heating temperature, welding conditions and post-weld treatment.

In cases where a repair weld must be made, a thorough analysis is required to determine whether the repair weld can be carried out successfully in situ or whether it must be made in the workshop or by an experienced foundry. In making this appraisal the following questions must be answered:

- whether adequate access is available in situ to detect all relevant cracks, to make the weld and to carry out post-weld heat treatment,
- whether the cross section to be welded can be adequately stress-relieved by local heat treatment,
- whether the resulting distortion of the casing due to welding and local heat treatment will be detrimental to the casing and
- whether the residual stresses caused by local heat treatment will reduce the safety of the component against brittle failure and its performance. Its performance, for instance, could be unduly impaired by the relaxation of the residual stresses at operating temperatures and the resulting distortion.

Good results have been obtained in connection with in situ weld repairs for relatively straightforward geometries of valve chests with thin wall thicknesses. In the case of thicker cross sections it is sometimes possible to achieve the target temperature of post-weld heat treatment by additional heating from the inside, provided the geometry of the interior permits the attachment of appropriate heating elements.

Where local post-weld heat treatment can only be carried out by heating from the outside, it is important to know that for each 1 mm wall thickness there is a temperature drop of at least 1 °C from outside to inside, as determined by optimum post-weld heat treatment of pipe welds (Cress 1967). This temperature difference should be compared with the permissible stress-relieving margin for the individual cast steel grades.

According to DIN 17 245 welded joints of 1% CrMoV cast steel, for instance, call for a minimum temperature of 680 °C at a maximum permissible stress-relieving temperature of 740 °C. Any appreciable shortfall in the minimum stress-relieving temperature of 700 °C results in embrittlement under creep stressing and in general time-dependent embrittlement (Cress 1967).

This applies in particular to the heat affected zone which during welding is exposed to a peak temperature up to about 1300 °C. If, on the other hand, the upper range of 740 °C is exceeded, the A_{c1} temperature will be exceeded and will result in an intercritical microstructure with a low creep strength.

An in situ repair is practically ruled out in cases where complex welding of valve chests with a complicated geometry is necessary. For such valves it is recommended to have the repair carried out by a qualified foundry which is equipped with the necessary inspection and fabrication facilities, ensuring that all crack zones are discovered, that adequate access is available to the crack zones for welding, and that stress-relieving can be carried out in a furnace after welding. The foundry would also be capable of machining the component to correct any distortion which might have occurred.

Turbine casings are less suitable for in situ weld repair because the upper and lower halves have to be separated and, as a result, local preheating, welding and local post-weld heat treatment are liable to result in undue distortion.

Furthermore, local post-weld heat treatment tends to release further residual stresses which in the long term relax at the operating temperatures, thereby producing distortion which can result in leaking joints and other losses of satisfactory performance. Here, too, good experience has been made with repair welds by the foundry. By following this course, a satisfactory and dependable repair is achieved in consideration of all the above aspects. An apparent time and cost benefit of in situ repair is partly negated by uncertainty in further operation and by extensive additional work.

Figure 9 shows the upper and lower halves of an HP outer casing pre-stressed by a system of shrink-fit rods for welding and heat treatment. To avoid scaling during heat treatment in the furnace at 690 °C, the casing was placed in an argon-flushed tank. As a result of the restraint, the distortion of the joints was limited to a maximum of 1.0 mm.

In the international literature there are also recommendations to use heat-resistant nickel alloy weld metals for local repairs on turbine casings and valve chests without preheating during welding and without subsequent post-weld heat treatment. The disadvantages of this method are:

- high hardness, low ductility and low toughness in the heat affected zone of the 1% CrMoV bare metal,

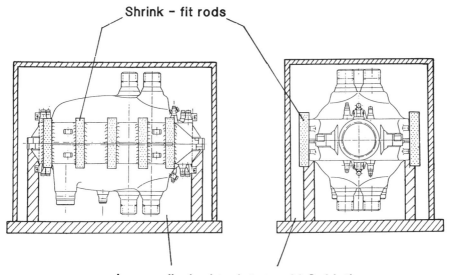

Shrink – fit rods

Argon – flushed tank to avoid Oxidation

Figure 9. Post-weld heat treatment of a repair-welded HP outer casing in a furnace (Argon atmosphere)

- long term embrittlement in the heat affected zone of the base metal,
- high local residual stresses in the welded area,
- relatively high local thermal stresses due to the difference in thermal expansion of the base and the weld metal.

By applying this repair recommendation there is a very high risk of cracking in the welded area.

SUMMARY AND CONCLUSION

The inspection of long-term-stressed steam turbines of fossil-fired power stations frequently brings forth cracks in the casings and valve chests of the HP and IP turbines. The cracks, as a rule, are located in regions of cross sectional transitions and at casting or weld defects. They are attributable to high thermal stresses occasioned by transient operating conditions.

Experience has shown that the casings - unless the cracks can be removed by grinding - can be reconditioned by weld repair. The question of whether the weld repair can be carried out in situ, or whether it should be carried out in the workshop or by a qualified foundry, must be decided on the merits of each individual case. In situ repair welds can be carried out successfully on straightforward geometries of valve chests with thin wall thicknesses. For valve chests with complex geometries and poor access to the regions to be welded and for large repair welds of turbine casings, it is only in a qualified foundry or workshop that a dependable repair can be guaranteed.

REFERENCES

Bareiss, J. & Maile, K. & Berger, G. & Mayer, K.H, and Weiss, M.
Beschreibung des Rißeinleitungs- und Rißfortschritts-verhaltens von Warmrissen in warmfesten 1% CrMoV Gußstücken unter Kriech- und Ermüdungsbeanspruchung, 17. Vortragsveranstaltung Langzeitverhalten warmfester Stähle und Hochtemperaturwerkstoffe, Düsseldorf, Germany, 1994.

Brühl, F.R. & Müller, H.H.
Beeinflussung der Temperaturverteilung bei der induktiven Wärmebehandlung von Hochdruckrohrleitungen aus warmfesten Stählen, Schweißen und Schneiden 1972, Heft 1.

Christianus, D. & Prestel, W. & Bächthold, H.
Ultraschallprüfung von Stahlguß nach DIN 1690, Teil 2 und SEP 1922 1985, Konstruieren und Gießen 12, 1987, Heft 2, Pages 19-30

Cress, T.
Zum Zeitstandverhalten warmfester Chrom-Molybdän-Vanadium-Stähle und deren Neigung zu verformungslosen Brüchen, Dissertation Technische Hochschule Darmstadt, Nr. D17, 1967.

Mayer, K.H. & Gysel, W. & Trautwein, A. & Tremmel, D.
Die Anwendung des warmfesten Stahlgusses im Dampfturbinenbau - seine Anforderungen und Eigenschaften. VGB Kraftwerkstechnik 60, 1980, Heft 5, Seiten 398-405.

Woodcock, S.J.
Upgrading opportunities for high temperature steam turbines. EPRI Seminar on fossil plant retrofits for improved heat rate and availability, San Diego, Ca., USA, Dec. 1-3, 1987.

Ageing of Materials and Methods for the Assessment of Lifetimes of Engineering Plant, Penny (ed.)
© *1997 Balkema, Rotterdam, ISBN 90 5410 874 6*

Planning of power plant service and rehabilitation work
Reasons for service life extension of older plants

H. R. Kautz
Grosskraftwerk Mannheim AG, Germany

ABSTRACT: For various reasons - in Germany it is e.g. lacking acceptance - the construction of new power plants is becoming more and more difficult. However, there are also other reasons than the personal uncomfort. Fig. 1 illustrates the changes in the management of US utilities over the past three decades. As a result of the installation of commissions, decision making is getting more complex. The attitude towards risks has changed, competition became tougher, financing is also difficult meanwhile. Environmental problems were unknown 30 years ago; the size of a plant could not be big enough! In the USA, high interest rates and increasing construction costs force the utilities to retrofit also small power plants with a capacity of 60 MW and more and to continue operation (Ullmann 1990).

	Sixties	**Nineties**
decision making	by utility management	more and more by commissions
assessment of risk	risks are small and negligible	risks are high, must be prevented
competitive pressure	negligible	by industrial power plants
financing	occasionally required	central problem
environment	secondary problem	decisive topic
size	bigger is better	small to medium size is realistic

Figure 1. Changes in operations management at US Utilities (Ullmann 1990)

1 POWER PLANT COMPONENT FAILURES DUE TO CORROSION/FIRESIDE CORROSION, AGING, CREEP, LOW-CYCLE FATIGUE

Figure 2 lists the various loads to which the many components of a power plant are subjected and which failures occur.

For example, flue gas desulfurization plants (FDG) - here the ammonia storage vessels - are stress-loaded under ambient temperature. Corrosion affects the end of life of steel structures and concrete constructions. In the past, ammonia storage vessels (for $DeNo_x$ systems) frequently experienced failure as a result of stress

Pipe Lines and Vessels in the Hot Yield Range	
• feedwater pressurized lines • steam extraction lines • cold reheat lines	• boiler drums • deaerator storage vessel • high-pressure preheaters ....corrosion/erosion/thermal stresses

Components Operated in the Creep Range

boiler (steam generator)
components affected by operating time/temperature/corrosion

• headers	... creep/cycling strain fatigue
• isolation valve assemblies	... creep/cycling strain fatigue
• steam lines with forgings and bends	... creep strain (weld joints)
• steam generator wall/economizer	...creep/cycling strain fatigue/corrosion

turbine-generator
components affected by operating time/temperature/corrosion:

• steam chest/isolation valve assemblies	thermal cycling strain fatigue with elevated temperature - creep (weld joints)
• cycle pipe lines	... creep strain (weld joints)
• turbine casing	... cycling strain fatigue (weld joints)
• turbine rotor	... creep strain/fatigue

Components Operated at Ambient Temperature

civil engineering components
components affected by aging/environment:

• external structures	... corrosion

flue gas cleaning systems

• ammonia storage tank	... stress corrosion cracking

Figure 2. Loads and failures

corrosion cracking in the weld area. Coal pulverizers are subject to severe wastage as a function of the coal ash content.

Many pipe lines, e.g. the cold reheater lines (CRL), but also many vessels, such as boiler drums, feedwater tanks, and the preheaters are operated in the hot yield range. They are subject to corrosion erosion, frequently also wastage.

The components operated in the creep range, such as steam generator, turbine-generator, high-pressure pipe lines are subject to aging due to creep, cycling fatigue, fireside corrosion etc.

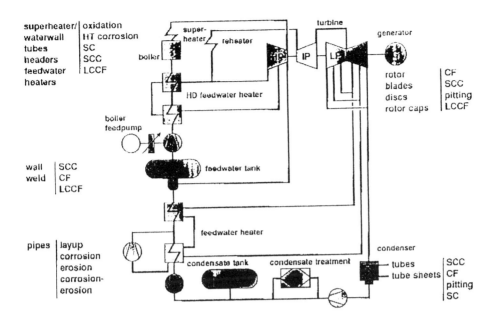

Figure 3. Corrosion of plant cycle components.

Figure 3 is a survey of the (possible) corrosion of plant cycle components. The steam generator (superheater, waterwalls, headers, preheaters) experience oxidation, fireside corrosion, strain-induced corrosion cracking, stress corrosion cracking, low-cycle corrosion fatigue (Hagn 1992).

In order to determine the condition of the microstructure and the degree of material exhaustion more and more long term component metallography analyses are conducted for plant components operated in the creep range. These analyses are based on the Outline "Steam Generators" of the German Association of Supervisory Organizations No. 451-83/6: "Surface Microstructure Examination of Creep-Stresses Components Pursuant to German Boiler Code TRD 508" and on the VGB[*] Guideline R 509 L "Recurrent Examinations of Pipe Line Systems in Fossil-Fired Steam Plants". The "Assessment Classes for Evaluating Microstructure Condition and Damage in Creep-Stresses Materials of High-Pressure Lines and Boiler Components" prepared by the VGB* almost four years ago represent a summary of the available documentation, information and findings so far. They are intended for an improved cooperation between the parties concerned. Figure 4 is a survey of the microstructure damage and the assessment classes. In agreement with an extended classification used for many years the classes listed in Fig. 4 are established. For determination of creep damage the surface microstructure analysis is the safest method!

---

[*] German Association of Power Plant Utilities

169

Assessment class	Structural and Damage Conditions
0	as received, without thermal service load
1	creep exposed, without cavities
2a	advanced creep exposure, isolated cavities
2b	more advanced creep exposure, numerous cavities without preferred orientation
3a	creep damage, numerous oriented cavities
3b	advance creep damage, chains of cavities and/or grain boundary separations
4	advanced creep damage, microcracks
5	large creep damage, macrocracks

Figure 4. Assessment classes for material microstructure

2 CONDITION ASSESSMENT AS A PREREQUISITE FOR REHABILITATION AND SUBSEQUENT MAINTENANCE

It cannot be repeated often enough that the prerequisite for any rehabilitation and subsequent maintenance presupposes an accurate condition assessment of steam generator, pipe lines, turbines and the vital auxiliary systems.

During the VGB "Service Life" conference at Mannheim in 1992 Barry Dooley and coworkers of the electric power Research Institute (EPRI) presented a remarkable paper on "Life Extension and Component Assessment in the United States" describing the so-called 'three level approach' to condition assessment of high-energy piping, a method applied to various areas at hierarchic levels of a power plant (thus also referring to the necessary and varied qualification of the staff). Below, this approach will be described briefly (Dooley et al 1992).

In the US, the assessment of aging fossil-fired power plants in the early to mid-1980's often involved major outages (three months minimum length) during which the condition of literally hundreds of pieces of plant equipment were examined, many through the use of expensive (and often redundant) sampling programs Often there was no balance in the approach with very detailed evaluations of one component and little or no evaluation of equally critical components elsewhere in the unit.

As a result of the shortcomings in such methods a program was developed to formalize and optimize for US utilities an approach to be taken for life extension. After the development of generic guidelines, a demonstration project was put in plate with ten US and Canadian utilities to test the general applicability of the methods. The general approach for an overall life optimization strategy has been extensively documented. Central to the assessment of particular equipment is a progressive, multi-level approach, as philosophy pioneered at the CEGB for the evaluation of high-temperature headers. Fig. 5 shows the defining characteristics of each of three levels of analysis.

Feature	Level I	Level II	Level III
failure history	plant records	plant records	plant records
dimensions	design or nominal	measured or nominal	measured
condition	records or nominal	inspection	detailed inspection
temperature and pressure	design or operational	operational or measured	measured
stresses	design or operational	simple calculation	refined analysis
material properties	minimum	minimum	actual material
material samples required?	no	no	yes

Figure 5. Data requirements for a multi-level approach (Dooley et al 1992).

The multi-step approach to the assessment has evolved and has been adopted by most US utilities. The approach is modified when common sense dictates that activities generally applied later in the sequence need to be moved to earlier roles. For example, it is often expeditious to identify critical locations or to determine allowable flaw sizes analytically before embarking on an inspection program (Level II) even though such activities are generally considered to be part of the most detailed evaluations (Level III). Such "pre- analysis" is most clearly advantageous when the analysis of inspection indications is complex and likely to take longer than that allowed by the constraints of the outage, or where knowing the critical damage level can trigger a replace-or-repair option.

As a result of the use of phased methods and multi-level equipment assessment, life extension activities have been matched to normal utility outage with concurrent cost savings.

Methods for Assessing Creep Damage

Methods to detect developing creep damage and to predict remaining life of high-temperature components once creep cracks initiate have been under extensive development during the past few years. Viswanathan and Gehl (EPRI Palo Alto, USA) have reviewed advance in life assessment techniques for these components including methods for detection of creep damage and for creep crack growth. Figure 6 overviews the available methods including open issues, techniques in bold print were specifically polled in the survey. Figure 7 provides a list of the applications and some additional notes for the key techniques.

The compilation of data should be organized right from the start in such a way that a transition to a computer-aided maintenance is possible at a later date. The goal is an integrated process control system. Figure 8 shows the gamut of tasks of the technical operation management. Figure 9 illustrates the concept of a modular operation control system (MOCS). The MOCS satisfies the tasks of an information system and supports the operational activities by way of informative and operative user's software modules.

Crack Initiation Technique	Crack Initiation Issues	Crack Propagation
Microstructural evaluation	Quantitative relationships with remaining life are lacking	
Cavitation measurement Carbide-coarsening measurements Lattice parameter Ferrite chemistry analysis Hardness monitoring		
Oxide scale measurements for tubes	Kinetics of hot-corrosion and constant-damage curves	
Replication strain monitoring		

Figure 6. Life assessment techniques and their limitations for creep damage (adapted from Viswanathan and Dooley, EPRI.

Crack Initiation Technique	Crack Initiation Issues	Crack Propagation
Calculation	Inaccurate	Technique: Analysis combines NDE, stress analysis and crack growth with an end-of life criterion to predict remaining life
Extrapolation of past experience	Inaccurate	
Conventional NDE	Inadequate resolution	
High-resolution NDE: Acoustic emission Positron annihilation Barkhausen noise analysis	Not sufficiently developed at this time	Issues for crack propagation methods:
	Uncertainty regarding original dimensions	Uncertainties in NDE results
Strain (dimension) measurement	Lack of clear-cut failure criteria	Lack of adequate crack growth data in creep and creep-fatigue
	Difficulty in detecting localized damage	
	Difficulty in sample removal	Lack of clear-cut end-of-life criterion under creep conditions
	Difficulty in using as a monitoring technique	
	Validity of life-fraction rule	Difficulty in assessing toughness of in-service components
Rupture testing	Effects of oxidation and specimen size	
	Uniaxial-to-multiaxial correlations	

Figure 7. Life Assessment techniques and their limitations for creep damage (adapted from Viswanathan and Dooley.

172

Process Management	Operation Management	Business Administration
process control	plant documentation	cost method
process monitoring	maintenance	performance of contracts
recording of accidents	material management	accounting
elimination of accidents	personnel management	personnel administration
release for operation	occupational safety/radiation protection	
	reports, statistics	
process control system $\Leftarrow$	Δ $\Leftarrow$operations control system$\Rightarrow$	$\Rightarrow$business administration system
	computer-aided	

Figure 8. Survey of the task of the different administrative plant sections (Wurzer 1989).

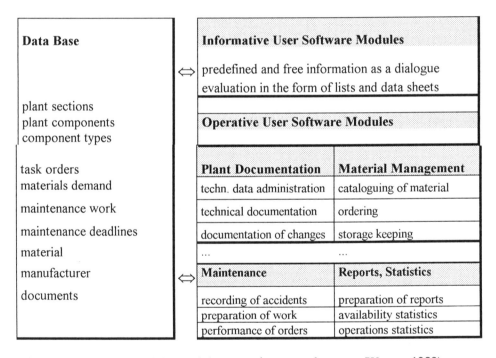

Figure 9. The concept of the modular operation control system (Wurzer 1989)

For older plants the question is frequently: Is a rehabilitation profitable? Are there replacements available? Is the installation of modern instrumentation and control profitable? Instrumentation and control, regardless whether coal-based or nuclear power generation, plays a critical part. West European countries today reach a service life of 40 years and more, with an upward tendency. Instrumentation and control contributes substantially to an extended service life because it allows an operation saving on the service life. Thus the plant life may be considerably extended by installing or modernizing I & C. So the installation of modern I & C should be an integral part of power plant rehabilitation.

3 REHABILITATION OF POWER PLANTS

A rehabilitation may be based on different aspects. In East Europe emphasis will be on the goal of generating "more energy", while in East Germany, after 1900, as a result of legislation, the decision was in the direction "reduction environmental pollution". Decisive is the age of the power plant, the capacity, the system's condition. The difficulties in obtaining replacement parts was already mentioned. Figure 10 is a list of criteria for selecting a maintenance approach, such as value of a plant (because of the geographical location), failure frequency of the whole plant or individual components, hazard potential for man and environment, procurement of third party power and costs of generation loss up to the consequential costs and damage. Figure 11 summarizes the tasks of boiler rehabilitation. The stress is on retrofits and upgrading of the boiler system. Upgrading of lignite-fired steam generators in East Germany aims at reducing the NO_x emissions by combustion process changes, furthermore, at increasing the boiler efficiency by modification and at service life extension, e.g. by replacing older pressurized parts and by adapting assemblies to the new legislation.

```
CRITERIA FOR THE SELECTION OF A
        MAINTENANCE CONCEPT

    • value of the plant
    • failure frequency
    • effect of failures on service life
    • hazard potential for man and environment
    • costs of loss of production
    • consecutive damage and costs
```

Figure 10. List of selection criteria

4 EFFICIENCY IMPROVEMENT BY REHABILITATION AND MAINTENANCE

In the past years, the crucial item for German utilities was the requirement of retrofitting existing plants or those under construction with flue gas cleanup systems, FGD and $DeNO_x$. Most of the plants retrofitting in this way were in operation for many years already. In the past years, more requirements were added, such as the improvement of the efficiency of operating plants in order to reduce the CO_2

emissions. Especially the public awareness which intensified over the past years enforces upgrading of the fossil-fired plants in order to improve their efficiency. Measures increasing the efficiency of power plants are:

- elevation of the main steam parameters,
- use of (double) reheating,
- multi-stage feedwater heating,
- air preheating,
- reduction of turbine losses,
- optimization of blade profiles
- decrease of the waste gas temperature,
- upgrading of the combustion process.

Many of these measures may be adopted only when erecting new plants. However, there is a number of measures which may be realized in plants already in operation. Figure 11 illustrates the improvement potential (not only of operating plants) by modifications at the turbine stages and the cold end and with respect to components, such as the steam generator. Repair welding of blades erosion due to droplet impingement up to optimization of blade profiles of the high-pressure, intermediate and low-pressure turbine stages. Here considerable progress was achieved in the development of upgraded blade profiles. The efficiency improvement starts at the high-pressure, intermediate pressure and low pressure turbine stages (HP, IP, LP), continues at the cold end to the coal drying system and ends with measures concerning the steam generator down to the topping gas turbine. Figure 11 shows also the efficiency of new plants just for power generation based on coal and dual-fuel systems with 30% natural gas. A topping gas turbine contributes considerably to the efficiency improvement. Figure 12 shows the technical data prior to and after the implementation of such measures. An improvement of the combustion by installing low-NO$_x$ burners is also feasible.

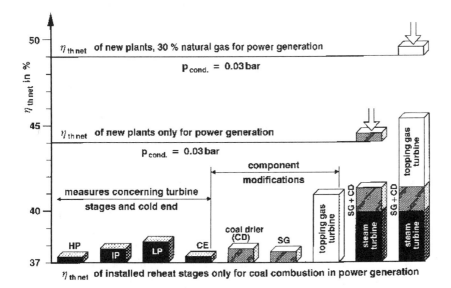

Figure 11. Improvement potential for older power plants (ABB Kraftwerke AG)

175

		Prior to modification	After Modification
steam turbine capacity	[MW]	511	473
gas turbine capacity	[MW]	-	135
total capacity	[MW]	511	608
capacity difference	[MW]	-	97
main steam pressure	[bar]	180	161
main steam temperature	[°C]	535	535
temperature behind reheater	[°C]	535	535
boiler fuel		gas/oil	gas/oil
gas turbine fuel			gas
fuel energy	[MJ/s]	1237	1325
energy differential	[MJ/s]		88
net thermal efficiency	[%]	41.3	45.9
efficiency increase	[%]		11.0

Figure 12. Technical parameters prior to and after modification (ABB Kraftwerke AG)

5 MAINTENANCE AND REMAINING LIFE - CHANGES OF MAINTENANCE PHILOSOPHY

Until well into the fifties, maintenance did not mean anything different than a repair, replacement of parts and return of the system to service.

At the beginning of the fifties, boiler capacity in West Germany increased rapidly. This increase brought about a change in plant maintenance philosophy, predominantly with respect to the steam generator and piping. On then the importance of availability and failure statistics as a support for decision making in plant maintenance was acknowledged. Around 1970 there are first indications to this. Strategies were developed with the objective of upgrading the reliability of power plants. The tendency to increase the unit size revealed the failure risk connected with such systems and, at the same time, enhanced the general acceptance of maintenance. This basic change results in the "maintenance philosophy" valid today, meaning monitoring, examination, renewal, upgrading, prevention. Decisive is the organization of information transmittal. Plant life extension became a complex planning goal. Figure 13 illustrates this change starting with the operation up to the failure of a component, the transition to preventive maintenance and finally to a condition-based maintenance. The three types of maintenance are used today in power plants. For failures (occurrence of upset conditions) redundancy is required. In other words, the plant components concerned must not have any impact on the availability, and there must be no risk of consecutive failures. In the second case, the components are repaired at established intervals. In the third case, monitoring methods are required for any early detection of incipient damage. A number of methods were successfully developed which allow in-process online monitoring. During short

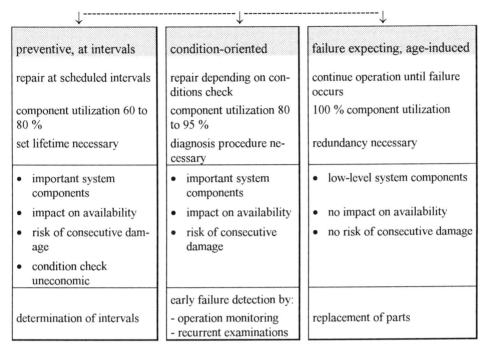

Figure 13. Maintenance strategies

- Adequate (physical) variables are continuously or periodically recorded and compared to target values

- Excursion of target values results in an alert or alarm message. This alarm indicates a system condition inadequate to operating conditions.

- An event analysis may be based on stored data. An extended diagnosis is possible, if an adequate knowledge base is available. This will allow a prediction of the further system performance and advice as to staff actions.

Figure 14. Tasks and functions of monitoring and diagnosis systems

outages or regular examinations during overhauls a documentation is prepared. This is the basis for the condition-based maintenance. Below is a short description of the tasks and functions of monitoring and diagnosis systems (e.g. life time monitoring systems - Figure 14).

The reason for discussing the service life of power plants is that the boundary conditions for the erection and operation of modern fossil-fired power plants have

177

Spherical Forging for a Hot Reheater

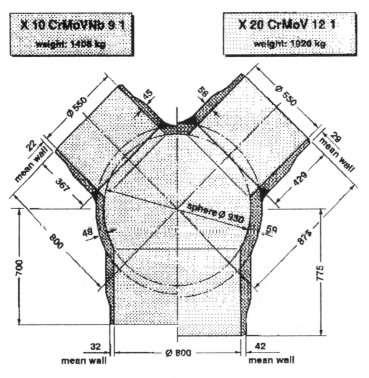

Figure 15. Comparison of material saving in a spherical forging.

greatly changed (deteriorated) over the past years. The existing power plants must be operated longer than planned after the retrofits with FGD systems. Another reason for retrofitting operating plants is the transition of the operation of older plants from baseload to cycling or even peak load operation.

6 MATERIALS FOR MAINTENANCE - WELDING AND MAINTENANCE

The use of the already mentioned steel X10 CrMoVNb 91 (P 91) allowed to raise the main steam temperature by approx. 20K as compared to the steel X20 CrMoV 12 1, however a still greater increase of the reheat temperature. The wall thickness and the weight may be reduced considerably as compared to X20 CrMoC 12 1, as may be seen from Figure 15 - a spherical forging for a hot reheat line for 565°C. However, decisive is the fact that the steel X10 CrMoVNb 9 1 has improved welding characteristics as compared to X20 CrMoC 12 1 which, together with its elevated high-temperature strength, stressed its importance for the construction of power plants, but also for maintenance work.

Finally, a very unusual rehabilitation by welding shall be described. Not only in Germany, but also in England (National Power) and South Africa ESKOM) the lines made of MoV were rehabilitated by welding with nickel-base fillers without subsequent heat treatment (Brett). The standard procedure - similar welds and subsequent stress relief annealing - resulted in cracks in the heat-affected zone so that the only solution left was to repair without subsequently annealing by using a nickel-base filler.

CONCLUSION

For various reasons - in Germany the lacking public acceptance - the construction of new power plants becomes more and more difficult. However, there are still other reasons than personal uncomfort. New plants demand a great amount of capital. Licensing procedures aggravate the decision for a new plant. The new concepts for lowering power generating costs have not yet reached commercial standard. The financial situation of a utility is frequently a difficult one. Last, but not least, the social policy in the utility environment is one of the decisive factors today. The installation of commissions results in an aggravation of decision making for a new plant. The attitude with respect to the operational risks has changed, the pressure by competitive producers intensified. The financial difficulties are not in the least caused by the fact that frequently smaller sizes are recommended, while a higher plant capacity would be much more profitable. The difference between a 300 MW and a 1000 MW unit may run to 25%.

Thirty years ago environment problems were unknown. Plants could never be big enough! All this forced utilities, e.g. the USA to consider upgrading and operating smaller plants - starting with 60 MW, especially in urban areas - for another 50 to 60 years.

REFERENCES

ABB Kraftwerke AG, N.N. "Retrofitting for improving the efficiency of fossil-fired power plants" (in German). Issue DKW 602590 D.

Brett, S. "Methodology for the use of weld repairs without power-weld heat treatment on creep resisting steels". BALTICA III, Vol. II, pp 363-373

Dooley, R.B. McNaughton W.P. and Viswanathan R. "Life extension and component condition assessment in the United States". Proc. Int. Conference at Mannheim, 1992.

Hagn, L. "Possibilities and limitations of service life calculations of steam generators and turbines" (in German). *Maschinenschaden 65* (1992) pp 117-126.

Ullmann, K. "Technical and political perspectives of power generation". *VGB Kraftwerkstechnik 70* (1990), pp 102-105.

Wurzer, K. "Software for operating power plants" (in German). Siemens Service Report No. 10 (1989), pp 8-10.

© 1997 Balkema, Rotterdam, ISBN 90 5410 874 6

Life assessment methods for low alloy drum steels after long-term service

A. Hernas
Department of Materials Science, Silesian University of Technology, Katowice, Poland

L. Mirecki
Boiler Factory 'Rafako' SA, Racibórz, Poland

ABSTRACT: One of the essential metal science problems related to the long-term service of pressure components of the power installations is a necessity of evaluation of their residual life time. The paper presents an approach to the extension of safe service life, application of versatile methods for testing and measurement of boiler drums after long-term service. Non-destructive test results, including metallographic inspection are used as the starting point for decisions with regard to destructive testing of the drums. Some results of investigation of microstructure (using SEM and TEM), mechanical properties and fracture surfaces of low alloy copper-containing steel after long term service and restoration heat treatment is described. After about 180 000 hr service impact strength and fracture toughness (K_{IC}, δ_C are significantly low. Fracture toughness resistance of weldments is better than that of the base metal after long-term service.

1 INTRODUCTION

In the Polish power plants and thermal-electric power stations are operating different types of drums manufactured by various methods from carbon and low alloy steels (Table 1). The drums have diameters between 1300-1900 mm and wall thicknesses 55-100 mm; they operate at 320-350 °C in various types of boiler (e.g. PK10, PAUKER, OP-140, OP-230, OP-380, OP-650).

Table 1. Chemical compositions (in wt.%) and average values of standard yield strength (MPa) of boiler drums steels used in Poland.

Grade	C	Mn	Si	Ni	Mo	Cu (V, Al$_m$)	σ_y 20 °C	320 or 350 °C
18CuNMT	max. 0.20	0.7-1.0	0.3-0.5	1.0-1.2	0.25-0.32	0.90	min 397	304-451
CuNi52Mo	0.20	0.9	0.40	1.1	0.30	1.1	382	338
15NCuMNb	0.17	0.8-1.2	0.25-0.50	1.0-1.3	0.25-0.40	0.5-0.8	400	314
CuNi47	0.18	0.7	0.40	0.90	-	0.9	333	245
13 123.9	0.19	1.17	-	0.06	-	0.2 (V)	284	216
K22M	0.17-0.23	0.8-1.1	0.15-0.35	max. 0.3	0.25-0.40	0.25 (Al$_m$)	275	185
22K	0.29-0.26	0.7-1.0	0.17-0.40	-	-	-	265	186
St47K	0.14-0.34	min. 0.55	0.15-0.33	max. 0.3	-	max. 0.3	177	119

After over 30 year service carbon steel drums have in general satisfactory tensile properties and toughness in spite of the fact that their internal surfaces show advanced efects of corrosion, pitting and cracks. In those cases where the yield point or impact strength levels were below standard any rapairs were followed by annealing. On the other hand significant problems were encountered in the case of low alloy steel drums containing copper. Low alloy steels with copper and nickel additions of up to 1.2 wt. % exhibit high yield strength at 350 °C and good room temperature toughness. These steels show strong tendency toward hardening under thermo-mechanical cycling (Rau et al. 1992). Such problems occured in DIN 17CuNiMo4 (PN 18CuNMT) steel drums where inferior toughness of the material and high susceptibility to cracking in the weld metal and HAZ were common.

Changes in properties and hardening which control the life are caused mainly by microstructural processes: precipitation of copper and carbides or carbonitride particles, decomposition of pearlite/bainite areas, changes in carbide morphology and heterogeneity of ferrite grain size already in the state of delivery.

Boiler drums belong to the group of power components operating below creep threshold temperature (about 250-350 °C). They are designed on the basis of the minimum value of the elevated temperature yield strength, theoretically for the so-called unlimited life. However mechanical designe of boiler drums did not include nonstationary and variable loads which do occur during service. The operating conditions and especially the strain accumulation during cooling down and heating up cycles after each check up lead eventually to catastrophic failures.

The service life of components operating below the creep threshold temperature varies widely in spite of the fact that they are designed for unlimited time of operation. The components' durability may be affected by many factors related to design, manufacture and operation conditions. The life of drums is closely associated with destructive processes taking place during operation, namely: thermal and thermo-mechanical low cycle fatigue, corrosion, brittle fracture and service cracking. In real service conditions the life of the components is usually affected by some combined influence of these processes (Trzeszczyński 1993, Kaczorowski 1995).

2 METHODOLOGY OF THE ASSESMENT OF MATERIAL CONDITION.

Progressive depreciation and growing capital costs of pressure installations for power stations and petrochemical industry make the assessment of technical condition of the capability installations immensely important from technical and economic standpoint. Such assessment provides basis for decisions with regard to:
• fitness for continued safe operation in specified conditions,
• qualification for, and planning of repair, overhaul, modernization or replacement.
The complexity of the interactions between various factors and conditions, and the resulting changes in materials after elevated temperature service require a complex and reliable approach to the problem of extension of safe service life and application of non-destructive and destructive methods of testing, measurement and calculations. Every object should be treated individually, and the same is true for separate components of power installations (Dobosiewicz 1994). The approach to remanent life assessment is shown in Fig. 1.
In the first step of any remaining life assessment an analysis of the past operating

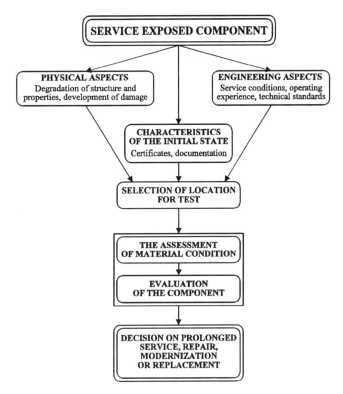

Figure 1. The approach to remanent life assessment of power plant component.

history is required, together with careful selection of potentially critical sites of pressure components. Selection of locations for tests is very important because of its decissive influence on the method and scope of the tests, the time needed, reliability of the decisions and costs. The selection is made on the basis of:
- knowledge and experience of the designers,
- knowledge and experience of the operators,
- certificates for the materials,
- data from previous tests and repaires,
- metallurgical knowledge.

The tests are usually made in sites of highest stress, inferior initial properties, unfavourable stress patterns. The most serious damage occurs mainly at the sites of stress concentration (Fig.2).

The program of tests is prepared individually for each drum. Over-all assessment of the object must be preceeded by structural recalculation, non-destructive and/or destructive evaluation of the material after service in specified conditions.

The non-destructive testing program for boiler drums used by "RAFAKO" S.A Service includes (Kaczorowski et al. 1995):
- external and internal visual inspection to detect e.g. of corrosion, erosion, deformation, damaged fastenings, etc.
- magnetic powder inspection of welds,

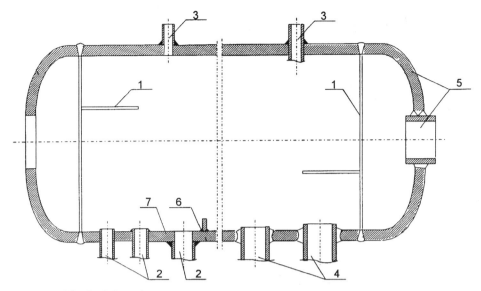

Figure 2. Typical locations of post-service cracks in boiler drums (Trzeszczyński 1993); 1 - longitudinal and girth welded joints, 2 - openings and their edges below water level, 3 - nozzles, 4 - downcomers, 5 - head and access sleeve, 6 - cracks under internal separation supports, found especially after dismantling, 7 - bridges between nozzles (Fig. 3),

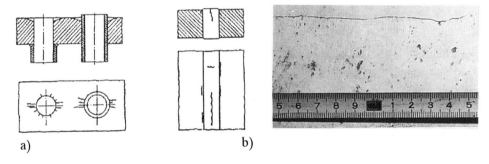

a)   b)

Figure 3. Locations of cracks between nozzles (a) and in the weldment (b).

- penetrant fluid tests for welds and internal surfaces of the openings for downcomers,
- ultrasonic inspection of welds, access sleeve, bridges; wall thickness measurements
- hardness tests and and geometric measurements,
- endoscopic inspection of internal surfaces of sleeve, gate and valve openings,
- measurement of damping intensity coefficient in drum ends (head),
- metallographic examinations with the use of the replica technique.

An analysis of the past operating history, material certificates and destructive testing results including metallographic data obtained by means of replicas and calculations are used as the starting point for decisions with regard to destructive testing, e.g.

184

drums are qualified for cutting out suitable test pieces (disks or rings cut out from places which were considered as the most dangerous from the standpoint of material life exhaustion). These test pieces are used for comprehensive mechanical testing, metallographic examinations and selection of restoration heat treatments.

Calculation methods are applied for a preliminary assessment, before more detailed examinations and measurements of the analysed object begin. These methods are also used for the recalculation of the drum material strength. In particular, the analysis of the status of stress under static load at service conditions is performed, followed by calculations or estimations concerning nucleation and propagation of cracks (critical and threshold crack depth, number of load cycles for transition of a crack from threshold to critical depth) and the analysis of low-cycle fatigue strength. The analysis of low-cycle fatigue strength is carried out according to the methods recommended by VGB, TRD and AD Merkblat and including:

- determination of the permissible number of hydraulic pressure tests,
- definition of the permissible number of shut-downs and start-ups,
- definition of the number of load changes at emergency conditions,
- determination of the total degree of exhaustion of the fatigue life.

In most cases strength calculations are based on estimated material data. Investigations of real mechanical properties of material after long term operating of drum are to be used for verification of the calculations in order to improve the reliability of decisions concerning the time of further safe service.

3 SCOPE OF REPAIRS AND MODERNIZATION

The metallographic assessment of the material status and other diagnostic tests, together with the analysis of recalculations for actual and design parameters, lead to final estimation of the component status. Then a decision can be made with regard to the permissible time of extended service in particular conditions or any applicable repair, or - in extreme cases - to replacement. Following the principle that the assessment of a component is based on the combination of methods, the rule is observed that fitness for additional service is based on the least favourable result among the methods used. According to the type of damage, degree of structural degradation, results of mechanical tests and structural calculations, information on previous repairs, grade of material and condition of fastenings, individual repair recommendations are prepared for each drum. These must be approved by the appropriate Technical Supervision Office (Kaczorowski et al. 1995).

The following methods of repair are used for boiler drums (Rau et al. 1992, Trzeszczyński 1993, Kaczorowski et al. 1995):

- removal of defects by grinding,
- deposition welding to restore material removed by grinding,
- repairs by welding of e.g. internals with preheating or sometimes without preheating,
- repairs of drum shells on edges, in openings and fittings by welding with preheating and so-called "thermal breath",
- deposition welding of drum shells to restore the required thickness,
- repairs by welding of drum shells and openings with local or over-all annealing.

Almost all repairs involve modernization of any outdated or defective design of drum shells and internals using up-to-date design principles, new welding techniques and technology used by RAFAKO S.A.

Results and analysis make it possible to:

- estimate the time of additional safe service of the components
- prepare recommendations with regard to service conditions with special reference to the start-ups (hydraulic tests - in most cases at temperature 50 °C and 0.8 of service pressure).
- prepare recommendations with regard to supervision, monitoring of service parameters and the time of the next tests.

4 RESULTS OF EXPERIMENTS

The experiments were carried out on samples taken from a boiler drum with a wall thickness 85 mm made of low alloy 18CuNMT (DIN 17CuNiMo4) steel (Table 1). The drum worked at 350 °C and was scrapped after about 1.8×10^5 hours. The investigations included examinations of the microstructure by LM, SEM and TEM techniques equipped with Link Exl EDS attachment enabling chemical analysis, mechanical properties with fracture toughness and low cycles fatigue of the base metal, HAZ and weld and fracture surface analysis. Some of the samples were annealed for 4-8 hrs at 580-650 °C.

4.1 Microstructure

After long-term service the microstructure of the steel consisted of fine- and coarse-grained banded ferrite and degraded bainite/pearlite areas (Fig. 4, 5). Locally M_3C and M_7C_3 carbides occured on grain boundaries in the form of a network. The ferrite grains contained numerous small disc shape copper precipitates of size from 0.01 to 0.05 μm (Morgiel et al. 1996), inducing strain contrast in the surrounding matrix (Fig. 6). Strengthening of ferrite by copper particles was noticed also in degraded post-pearlite/bainite areas (Fig.7).

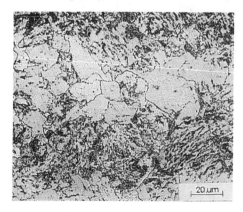 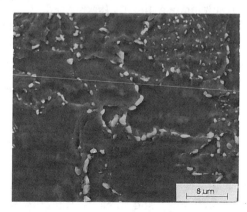

Figure 4. Banded ferrite-bainite/pearlite structure in 18CuNMT steel after 180 000 hr service.

Figure 5. Microstructural degradation of the 18CuNMT steel drum after service. SEM-metallographic section.

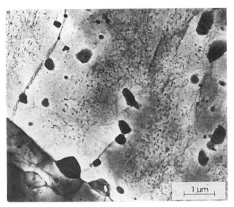

Figure 6. Dislocations and copper precipitates in the ferrite grain after service of drum. Thin foil.

Figure 7. Decomposition of pearlite/bainite area and copper precipitation in ferrite. Thin foil.

4.2 Mechanical properties

Mechanical properties of the steel in the state of delivery, after 140 000 and 180 000 hr service and also after restoration heat treatment are shown in Fig 8.

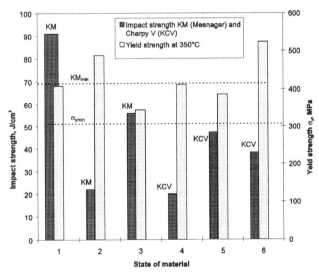

Figure 8. Changes in impact strength and yield strength at 350 °C after long term service and restoration heat treatment of 18CuNMT drum steel.
1 - delivery (initial) state, 2 - after 140 000 hrs of service, 3 - after heat treatment 640 °C/6 h., 4. after 180 000 hrs of service, 5 - after heat treatment 650 °C/4 h., 6 - after standard heat treatment of the material after 180 000 hrs of service.
Note: After 140 000 hrs of service the investigation were carried out on a disc taken from the drum and next the whole drum was heat treated (640 °C/ 6 h).

187

Low alloy drum steels containing copper suffer from structural and mechanical degradation resulting in significant losses of toughness as determined by impact tests and fracture/fatigue behaviour. The fracture tests were carried out according to BS7448: Part 1: 1991 using a MTS-system at room temperature and three-point bending test pieces with V-notch and thickness of 26 mm.

Average values from 5 tests of the crack extension resistance under conditions of crack-tip plane-strain K_{IC} and critical crack opening displacement δ_C of post service material with girth weldment is shown in Table 2. Average values of CTOD for material with longitudinal weldment after service and restoration heat treatment are listed in Table 3.

Ductility of the material with girth weld was worse than that of the material with longitudinal weld. This was obviously due to banding, anisotropy of mechanical properties and orientation of the notch with respect to the direction of rolling.

Table 2. Mechanical properties of material with girth weld at room temperature. The mean value of: critical crack openning (δ_c), stress intensity factor (K_{Ic}) and yield ratio σ_y/σ_{TS} and Charpy-V impact strength.

Parameter	Base metal	HAZ	Weld metal
δ_c, mm	0.020	0.047	0.155
K_{Ic}, MPA·m$^{1/2}$	49.12	63.11	135.33[a]
KCV, J/cm^2	12.50	20.30	46.80
σ_y/σ_{TS}	0.79	0.82	0.74

explanation: [a] - the value calculated from the equation: $K_{Ic} = [\sigma_y E \delta_c /(1-v^2)]^{1/2}$
(v - Poisson's ratio)

Table 3. The mean values of crack opening displacement δ_c, δ_u, δ_m [mm] of material with longitudinal weld.

Material 650 °C/4 hrs	after 180 000 hr service	after service and restoration heat treatment
Base metal	$\delta_c = 0.070$	$\delta_c = 0.112$
HAZ	$\delta_c = 0.380$	$\delta_c = 0.122$ and $\delta_u = 0.375$
Weld metal	$\delta_u = 0.312$	$\delta_u = 0.285$ and $\delta_u = 0.614$

Post-service low-cycle fatigue tests were carried out on the base material and on test pieces which included the base metal, HAZ and weld metal (Fig. 9). MTS testing machine was used with control strain at room temperature. Table 4 shows average results from three tests.

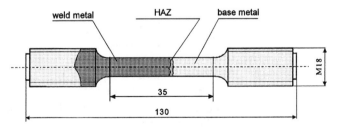

Figure 9. Low-cycle fatigue test piece.

Table 4. The results of low-cycle fatigue tests of material with longitudinal weld after service ($\Delta\varepsilon_{t,p,e}$ - mean value of three tests).

Sample	$\Delta\varepsilon_t$, %	$\Delta\varepsilon_p$, %	$\Delta\varepsilon_e$, %	N_f, cycles
With weld (Fig. 9)	0.78	0.36	0.42	1283, 1731, 2478
Base metal	0.79	0.33	0.46	1463, 1510, 1812

Yield ratio are exceded 0.72, indicating low resistance to brittle cracking. The lowest toughness value occured in the base metal (K_{Ic}<50 MPa·m$^{1/2}$ and impact strength KCV =20,5 J/cm^2).

Fractures of such materials after service are brittle in character (Fig. 10a). Impact transition temperature and the FATT lie under room temperature. Ductility can be restored by appropriate heat treatment (Kaczorowski et al. 1995, Hernas 1996) ensuring safe service of repaired drums. In this case the percentage of ductile fracture increased to some 40% at room temperature (Fig. 10b) and also the values of K_{IC} and δ_C increased especially for the base metal (Table 3).

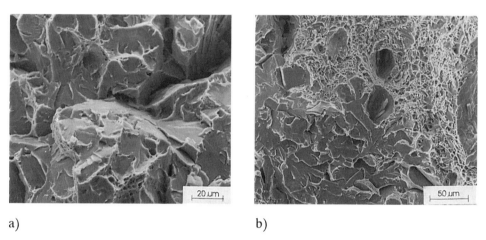

a)   b)

Figure 10. Brittle fracture of the material after long term service (a), brittle ductile fracture of the material after restoration heat treatment (b).

Standard heat treatment used for new drums failed to restore the initial toughness in post service drums (Fig. 8). In compare with the base metal and HAZ, the weld metal showed better ductility i.e. impact strength, crack resistance K_{Ic} and δ_C. This was contrary to practical experience which show that cracks in the weld metal and HAZ were the most frequent. On the other hand low-cycle fatigue tests showed that all welded test pieces cracked in the weld metal in accordance with practical experience. These results suggest that in considerations of cracking in boiler drums it is necessary to include threshold values of stress intensity factor and crack growth rate characteristics (Le May, 1995).

5 CONCLUSION

The most remarkable result was related to the role of copper in the strengthening of ferrite and degradation of bainite/pearlite areas of the drum steels (Fig. 4-7). It was found, that long-term service of 18CuNMT steel causes significant loss of toughness in particular the base metal, although initiation of fatigue cracks was most frequent in the weld metal. The mean value of impact strength (KCV), δ_C and K_{IC} were very low (Table 2-4), below minimum values required in standard specifications or in the literature concerning pressure vessels. Post-service restoration heat treatment at approximately 640 °C produces favourable results with respect to impact strength (Fig. 8) and the resistance of base material to fracture although the toughness of the HAZ and weld metal is hardly affected.

The results confirm once more that hydraulic tests of old boiler drums should be conducted at 50-60 °C i.e. above ductility transition temperature and at a pressure equal to 0.8 of service pressure. In such conditions of testing the risk of catastrophic is reduced.

The results will be used to verify current methods of structural design of boiler drums and will provide more reliable basis for evaluation of local loading of the component as well as forecasting of extended safe service time (Le May 1995) for boiler drums using steels containing appoximately 1% Cu.

During recent years RAFAKO S.A completed repairs and modernisation of more than 50 boiler drums of various types including 25 low alloy steel drums providing warranty for 30 000 hrs of safe operation at recommended service parameters.

REFERENCES

Dobosiewicz, J. 1994. *Energetyka* No 7, Biuletyn Pro Novum, No 1: p. 251 (in Polish).
Hernas A., Mirecki L. 1996. The microstructure and toughness of drum steels after long-term operating. *Proc. 8th Int. Symp. on Creep Resist. Met. Mat.*: Czech Republic.
Kaczorowski M., Dobrzański J., Hernas A. 1995. Comprehensive Evaluation of Material Condition in Pressure Components of Boiler and Prediction of Safe Residual Life, *Proc. II Sem. RAFAKO*: Warszawa (in Polish).
Le May I. 1995. Damage assessment and life extension in aging power plants, *Proc. FORUM'95 Central European and World Connection Electric Power Industry*: Cracow: p. 83-90.
Morgiel J., Szewczyk J., Baliga W. 1996. Microstructure changes in 17CuNiMo4 boiler plates, *Proc. IX Int. Conf. on Electron Microscopy of Solids*: Kraków-Zakopane, Poland: p. 507 .
Rau P., Schick M., Albert F., Helmrich A. 1992. *VGB Kraftwerkstechnik* 72, H.5: p. 444.
Trzeszczyński J., Zbroińska-Szczechura E. 1993. *Energetyka* No 3, p. 3 (in Polish).

Ageing effects

Under insulation corrosion – An inspection approach

A.W. Beattie
Inspection Department, Caltex Oil Refinery, Cape Town, South Africa

ABSTRACT: Under Insulation corrosion (U.I.C) is a real threat to the onstream reliability of many of petrochemical plants. This type of corrosion is difficult to detect as it is covered by insulation which masks the underlying corrosion problem until it is to late. The failures are often the result of localized corrosion and not general wasting of large areas. These failures can be catastrophic in nature or at least have an adverse economic effect in terms of downtime and repairs. Two common and costly examples, hidden from view under insulation, are non-uniform attack of plain carbon steels and stress corrosion cracking of austenitic stainless steel. Over the years, Calref has experienced numerous pipeline failures until a programme aimed at reducing the number of leaks caused by U.I.C. was implemented in 1988.

1 BACKGROUND

UIC takes place when water is allowed to enter an insulated system or component and contact the underlying surface. This water can originate as rain (UIC is predominant in high humidity, high rainfall coastal areas), run-off from equipment washdowns, deluge system tests condensation from temperature cycling and leakage from aqueous process systems. A minute leak from a steam tracer tube fitting or valve stem packing under insulation can cause major problems.

Insulation creates severe problems for inspection and maintenance and, consequently, corrosion can remain undetected until the piping has reached a critical condition. Leakage of hazardous chemicals often results and, in some cases, metal wastage has been so extensive that gross failure and major emissions have occurred.

By understanding the types of corrosion that can occur under insulation, appropriate materials and construction methods can be employed to prevent them. Effects of intruding water must be considered during the design to prevent corrosion which can be caused by by permitting water to enter a system directly or indirectly by capillary action.

Corrosion may attack the jacketing, the insulation hardware, or the underlying piping or equipment. Depending on other factors three types of corrosion may occur: galvanic; acidic or alkaline; and chloride stress corrosion cracking (SCC).

Galvanic Corrosion:

Galvanic corrosion generally results from wet insulation with an electrolyte or salt present that allows a current flow between dissimilar metals- i.e. the insulated metal surface and outer jacket or accessories. The extent and severity of the attack on the less noble metal depends not only on the difference in potential of the two metals, but also on their relative sizes.

Alkaline or acidic corrosion:

This mechanism takes place when an alkali or acid, and moisture, are present in certain fibrous or granular insulation's. For hot service above 150°C, most of the water is driven off. This water vapour may condense at the edge of the insulation, and dissolve the alkaline or acidic chemicals there, resulting in corrosion of the aluminium or steel jacketing.

Chloride SCC :

This type of corrosion is caused by the combination of insulation containing leachable chlorides with the 300 series stainless steel surfaces, whilst moisture is present at 60°C. Concentration of the chloride ion usually results from evaporation of rain water, or of water used to fight fires, or process water. Stress corrosion cracking of insulating jackets often results from airborne salts at coastal regions. The probability of failure and the speed of crack propagation are governed by the temperature of the stainless steel and the chloride concentration at the metal surface - as well as local stresses.

2 SUSCEPTIBLE AREAS

The American Petroleum Institute API 570, Inspection, Repair, Alteration and Rerating of In-service Piping Systems, the piping code first published in 1993 identifies U.I.C. as a special concern and identifies the following areas as susceptible:
   * Areas exposed to mist overspray from cooling towers.
   * Areas exposed to steam vents.
   * Areas exposed to deluge systems.
   * Areas subject to process spills, ingress of moisture, or acid vapours.
   * Carbon steel piping systems, included those insulated for personnel protection, operating between -4°C and 120°C. U.I.C. is particularly aggressive where operating temperatures cause frequent condensation and re-evaporation of atmospheric moisture.
   * Carbon steel piping that normally operates above 120°C but used in intermittent service.
   * Deadlegs and attachments that protrude from the insulated piping and operate at a temperature different from the active line.
   * Austenitic stainless steel piping systems that operate between 60°C and 204°C,

these systems are susceptible to chloride stress corrosion cracking.

* Vibrating piping systems that have a tendency to inflict damage to insulated piping providing a path for water ingress.

* Steam traced piping that may experience tracing leaks, especially at the tubing fittings beneath the insulation.

* Piping systems with deteriorated coatings and/or wrappings.

* Locations where insulation plugs have been removed to permit thickness gauging measurements on insulated piping should receive particular attention.

3 ALTERNATIVE INSPECTION METHODS

* *Profile Radiography* :

Exposures are made of small sections of the pipe wall. A comparator block such as a Ricki T is used to calculate the remaining wall thickness of the pipe. This method becomes technically challenging in piping systems over 10 inches in diameter.

* *Real time radiography* :

Fluoroscopy provides a clear view of pipes outside diameter through the insulation, producing a silhouette of the pipe on a TV type monitor that is viewed during the inspection.

The real time device has a source and image intensifier/detector connected to a C-arm. No film is used or developed. There are two main categories of RTR devices available today; one using a X-ray source and one using a radioactive source.

* *Infrared* :

In the right conditions infrared can be used to detect damp spots in the insulation, because there is usually a detectable temperature difference between dry and wet insulation. Corrosion is a distinct possibility in the areas beneath the wet insulation.

* *Neutron Backscatter*:

This system is designed to detect wet insulation on pipes and vessels. A radioactive source emits high energy neutrons into the insulation. If there is moisture in the insulation the hydrogen nuclei attenuate the energy of the neutrons. The count displayed to the inspector is proportional to the amount of water in the insulation.

* *Ultrasonic thickness gauging* :

This is an effective method, but limited to a small area. It is expensive to cut the insulation holes and cover the holes with caps and covers. It is not practical to cut sufficient holes to get a reliable result.

* *Insulation removal* :

The most effective method is to remove the insulation, check the surface condition of the pipe, and replace the insulation. This approach will detect CISCC in stainless steels and may require eddy current or liquid dye penetrant testing.

4 CORROSION HISTORY AT CALREF

Incidents at Calref :

* *Platformer Bypass Line:*

During 1987 a leak occurred on a bypass around a control station on the 2 inch platformer effluent line operating at 1650 kPa and normally 80°C. The bypass valve had been closed which effectively turned the bypass line into a dead-leg. The first sign of a problem was product seeping through the insulation, the insulation was removed and the problem diagnosed as U.I.C. This may have resulted in a serious failure but for the diligence of Operating staff.

* *Fractionator Top Pumparound Line:*

This leak occurred in August 1989, on the 10 inch bottom elbow of a vertical line on the crude fractionator top pumparound and resulted in a unit shutdown. The line operates at 90°C and at a low pressure of 720 kPa it is constructed from carbon steel with a nominal thickness of 6.4 mm. Examination of the line revealed water ingress through damaged insulation and subsequent corrosion.

* *B.F.W line to Sulphur Unit:*

This 2 inch line provided boiler feed water to the Sulphur unit steam generators. A leak occurred at a horizontal T-type support, which provided an entry point for moisture. Removal of insulation in this area revealed the corrosion to be isolated, though at other support locations to a lower severity. This line operated at a temperature of 126°C and a pressure of 1580 kPa.

* *Furnace Offgas Line:*

This is a steam traced 2" line feeding offgas to the furnace at an operating temperature of 124°C and a pressure of 60 KPa. Insulation was removed after gas smells were noticed in the area. Closer inspection revealed signs of sulphurous deposits escaping from the aluminium cladded joints. Minor steam tracing leakage had caused external corrosion which had been exacerbated by subsequent leakage of the H2S laden gas.

* De-ethaniser Column:

During a routine hydrotest after a vessel repair , a 3/4 inch nozzle on the vessel was found to be leaking due to external corrosion. The vessel operating temperature and pressure was 123°C and 1360 KPa respectively. The insulation adjacent to the nozzle wall has now been cut to within 100mm radius and the exposed area of the vessel wall and nozzle painted.

General

The incidents resulted in risk to personnel and/or lost production and enviromental pollution, all of which are unacceptable and require cost effective solutions. Several other leaks have occurred which have resulted in relatively minor inconvenience and expense to the Refinery, though which have served to re-inforce the need for action to prevent future occurrences.

5 INSULATION USED AT CALREF

The original Refinery construction used large quantities of asbestos insulation for both lines and vessel insulation. In order to mount an Inspection programme requiring the removal of this hazardous material, a procedure for its identification and safe removal where necessary, is required now by local legislation.

Newer Refinery construction projects use Rockwool. A high calcium, silicate, alumina material. This insulation is used in formed sections for pipe and blanket or board form for vessels. Aluminium jacketing is used extensively for both piping and vessels, though some use is made of finishing plaster for weather proofing.

6 ENVIROMENTAL CONDITIONS AT CALREF

The average Winter and Summer day time temperatures in the Cape Town area are 13.8°C and 19.5°C. The Relative Humidity taken at 14:00 for Winter and Summer is 58.5 and 52.8 respectively. Though generally not considered a highly corrosive area, strong Winter and Summer winds contribute corrosive salts for U.I.C.

7 U.I.C INSPECTION AT CALREF

Inspection Approach

Pipelines:

Central to the U.I.C system is the availability of accurate up to date Piping and Instrument Diagrams (P&ID`s), Isometrics and Flow Diagrams which accurately reflect the plant status, layout and provides operating information.

The procedure for the inspection and repairing and the lines was developed at Calref and is contained in the UIC Project Flow Chart (*Attachment A*) and is explained as follows:

The P & I D evaluation consists of the marking up of any lines that operate at temperatures at or below 150°C, and any lines that see intermittent use, such as steam out lines. The colour coding for the mark-ups is as follows:

Green - permanently remove insulation

Blue   - insulation has been removed, awaiting inspection

Orange - insulation to be removed for inspection and repairs, then replaced

For any lines that require permanent removal of insulation, a Permanent Insulation Removal (P.I.R.) request form must be submitted to Process Engineering Department, which will determine whether the permanent removal is viable or not. Once approval is granted the marked up P & I D's and isometrics are sent to the insulation contractor, and the insulation is removed as indicated in the diagrams.

The inspector will then examine visually the stripped lines and determine whether radiography is required. The radiography crew will perform the NDT on all the lines indicated by the inspector, and will submit the results. From the results, the inspector will issue QCP's for any repairs necessary, and for the replacement of insulation if required.

The repairs will be carried out by Reliability Services and any painting and/or insulation installation will be done by the insulation contractors.

For any permanent changes to the line, such as permanent removal of insulation or repairs differing from the original specifications, the P & I D's must be sent to the Drawing Office to be updated.

Vessels :

The U.I.C inspection of vessels is performed at the scheduled turnaround. Areas for consideration during the inspection are contained in a formal inspection guide which includes a checklist.

8 DATABASE

A database has been developed using Microsoft Excel, into which the following information is entered after each inspection :

* Line number
* The amount of insulation removed (meters)
* The date of the inspection
* The type of inspection used, visual, radiography, ultrasonics.
* The repair worklist number
* The inspection findings.
* The next inspection date

The above information and the link to the work order and costing will be used to cost and control future U.I.C inspection work.

9 DOCUMENTATION AND REPORTING

Wherever possible Calref has attempted to remove insulation permanently if not required for heat conservation. The formal identification, approval for removal and updating of P&ID`s has required the appropriate documentation integrated into the system.

All radiographs are identified together with the relevant piping Isometric drawing and stored for later retrieval and comparison.

The U.I.C Project is considered at Divisional Management level to be a priority issue, therefore a monthly status report of the "evaluation" and "action" work by plant is provided.

10 CALREF INSPECTION TECHNIQUES

Visual inspection and insulation removal is the primary technique for locating areas of corrosion or potential corrosion. The following are specific areas for inspection:

Lines:

* Leaking steam tracing.
* In the proximity of damaged insulation.
* Where insulation ends.
* At all breaks in insulation, e.g. at supports, valves, vents, drains, pipehanger supports, etc.
* Areas of personnel protection.
* Elbows at the bottom of a vertical run of pipe.

Vessels:

* Small bore nozzles on vessel shells.
* The top head of vertical columns operating below 150C. (i.e. cold end)
* The weather side of the vessel.(north side at Calref)
* At reinforcing and insulation support rings.
* On insulated tanks, the bottom 0.5m of insulation and at stairway and platform attachments.

Radiography:

Radiography is used to inspect lines insulated for heat conservation. It is applied after thorough visual inspection has taken place and potential areas for U.I.C. have been identified for further inspection. Should the line insulation appear in sound condition random areas for radiography are selected, typically adjacent to valves or where ingress of moisture could occur.

The advantages of using radiography over removing insulation are:

* Removal of insulation requires repair, which may induce additional areas for water ingress if not adequately repaired (i.e. more joints in insulation)

* Radiography provides a permanent record.

* Actual wall thickness measurements can be made.

* An estimate of internal condition of the pipe can be made.

Where situations occur and piping is discovered viaually as thin and which should not be disturbed, profile radiography is used to measure remaining wall thickness.

11 CONSTRUCTION AND MAINTENANCE

To ensure a high standard of insulation and protective finish the following points should be considered:

11.1 *Quality*

The application of the protective coating to the piping and the insulation, including its protective finish should be carried out by specialist contractors.

11.2 *"Completion date pressure"*

Since insulating activities are often carried out at the end of a project, increased pressure due to lack of time may have a detrimental effect on the quality of work performed.

11.3 *Repairs*

Repairs to damaged insulation should be performed as quickly as possible. Frequent inspection is therefore necessary to detect faults. In addition routine inspection of section of the piping should be carried out to check for corrosion.

Painting :

All piping inspected as part of the U.I.C. programme, operating at less than 150C, is painted. A paint system has been selected after various trials, which is tolerant of a low level of preparation. Sand or gritblasting in areas of on-line equipment is prohibited.

Drainage :

Water can be trapped in the insulation if provision is not made for draining through the protective finish. Metal clad systems may allow drainage via the lap joints whereas mastic covered finishes will have no in-built drainage unless poorly applied.

Unfortunately, the structure of most insulating materials increases the problem of effective drainage. Calcium silicate is particularly absorbent and closed cell insulation materials such as foam glass or polyurethane, though more resistant to moisture adsorption, can trap any moisture that penetrates a poorly applied jointing mastic.

General:

The design and maintenance of insulation will significantly contribute to the control of U.I.C. The following areas are worthy of consideration.

   * Does the equipment need insulation? If not, permanently remove.

   * Where insulation is required for personnel protection, the use of expanded aluminium can be used in place of normal fibrous insulation.

   * Design pressure vessel and piping attachments to prevent channelling of water into pockets.

   * Installation of weatherproofing such that joints are sealed, overlapped in the right direction.

   * Joints in horizontal runs of aluminium cladding are placed at the bottom, wherever possible.

   * After hydrotest of vessels, avoid soaking adjacent insulation by channeling/piping water away.

   * Holes are generally cut in insulation for ultrasonic gauging purposes, these should be firstly trepanned neatly and a custom made weatherproof plug installed.

   * The use of aluminium foil around steam traced lines to act as a vapour barrier, also used on insulated stainless steel lines to protect against stress corrosion cracking.

   * Quality Control is required on new insulation installations as part of construction inspection.

12 SUMMARY AND CONCLUSION

There is much insulated process piping and equipment inchemical plants and refineries which had not received immersion grade protective coatings before insulation was installed. The insulation systems were poorly designed/installed/ maintained and trouble in the form of corrosion and cracking is taking place. Although the latest procedures will never cover every single meter of piping installed they will reduce onstream leaks and hence lead to increasing plant reliability.

Attachment A - UIC Project Flow Chart
Attachment B - Photographs of U.I.C.

ATTACHMENT A

UIC Project Flow Chart

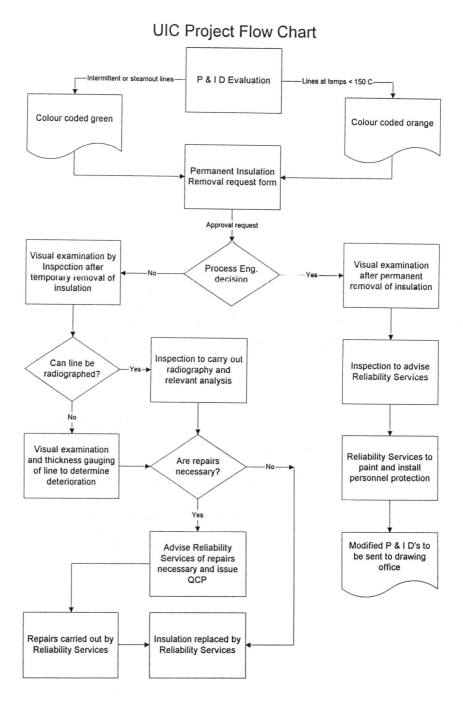

External corrosion of process line after personnel protection removed

Disintegration of emergency steam line to furnace. Line intermittently used

Corrosion of steam out point under insulation

Typical example of U.I.C. (crevice corrosion) found at a utility point.

External corrosion of pump by-pass line after insulation removed

REFERENCES

Korbin,G. 1995. "Corrosion under insulation". *Inspectioneering Journal*. Vol 1, TMSCo, California

Twomey, M. 1996. "Inspection techniques for detecting U.I.C". *Inspectioneering Journal*. Vol. 2, TMSCo, California

Mitchell L.V. 1987. "Preventing corrosion under insulation". Chemical Engineering Journal. SAIChemE, Johannesburg.

WWER-type nuclear reactor pressure vessel: Material radiation ageing issues and effect of thermal annealing as a mitigation method

V.I. Levit
Institute of Chemistry and Geosciences, Federal University of Pelotas, Brazil

ABSTRACT: Some degradation of the mechanical properties of WWER-type nuclear reactor pressure vessel materials (RPVMs) while in service of the reactor is one example of the radiation ageing (RA) of RPVM. The ageing mechanisms are complex and include the redistribution of impurity atoms, such as phosphorus, to grain and other boundaries and also the evolution of primary matrix damage caused by the high energy neutron irradiation. Many ageing mechanisms are influenced by thermally activated processes and the use of heat treatments at temperatures higher than where the ageing damage occurred provides a means to mitigate, at least partially, the adverse effects. The behaviour of radiation ageing of materials and recovery of their properties via annealing are analysed. Laws of radiation ageing of different materials have been considered. The case for irradiation-annealing and re-irradiation behaviour is also addressed. Analysis of destructive and non-destructive control methods for a given material state after irradiation and annealing is discussed.

1 INTRODUCTION

The assurance of the reactor pressure vessel's (RPV) reliable operation is one of the most important aspects in the safe operation of the nuclear power plant (NPP). The operational safety of the RPV depends, to a significant degree, on the resistance of the reactor pressure vessel materials (RPVM) to brittle failure under any operation conditions. The RPVMs are, from concept and manufacture, made from inherently tough, readily weldable ferritic steels and corrosion protection is usually given by the alloy composition or by a stainless steel cladding. Brittle failure of the RPV in a pressurised water reactor (PWR) or boiling water reactor (BWR) is to be avoided at all times since it is the primary pressure boundary of the coolant. Brittle failure is characterised by a rapid failure of the component without much prior indication. However, some degradation of the mechanical properties of RPVMs, while in service, is one example of the ageing of the RPVMs as the NPP accumulates a high fast neutron fluence over time at temperatures typically between 270 C-320°C. This degradation is generally termed as ageing. Radiation ageing (RA) consists of several significant mechanisms, but radiation embrittlement (RE) effects, which are characterised by an increase in the ductile-to-brittle transition temperature (DBTT) and also in radiation hardening (RH) and strengthening (RS) of the metal are, potentially, the greatest challenges to integrity. The ageing mechanisms are complex and include the redistribution of impurity atoms, such as phosphorus, to grain boundaries and other surfaces and also the evolution of primary matrix damage caused by the high energy neutron bombardment arising from fission processes in the fuel. Examples of the minute structural damage thus incurred are the production of higher concentrations of vacancies and the relative faster development of precipitates and a changing of their size distribution. All these changes will potentially affect the mechanical properties of WWER materials to varying degrees, depending also on factors such as neutron fluence, neutron energy spectrum of the damaging neutrons, material alloy composition (especially impurity elements such as phosphorus) and the irradiation temperature.

Many ageing mechanisms are influenced by thermally activated processes and the use of heat treatments (annealing) at temperatures higher than at where the ageing damage occurred provides a means to mitigate, at least partially, the adverse effects. Annealing of the vessel is one of the measures that has the potential to increase the operational life of the RPV which has reached the permissible (design allowance) embrittlement level. Using thermal annealing, the mechanical properties response of more than 10 reactor pressure materials in Russia and other countries have been examined.

The behaviour of radiation ageing of different RPVMs and the recovery of their properties due to applied annealing treatments are analysed in the following work.

2. RADIATION AGEING

Increase in the ductile-to-brittle transition temperature (DBTT) of RPVM during neutron irradiation is a very important effect of radiation ageing (RA) and it is usually called the "transition temperature shift" (TTS). Dependence of RA on chemical composition for different RPVM can be expressed as some empirical associations. For example, RPVM of WWER-type materials have some dependencies of TTS on Cu, P and Ni content (Figure 1 - 4).

The dependence of TTS of RPVM with low Ni content (Ni < 0.5 mass %) and mid-range P content (i.e. 0.010 mass% < P < 0.025 mass %) has a saturation at a Cu content near to 0.20 mass %. (see Fig. 1). The RPVM's TTS increases rapidly with Ni content when Ni exceeds 1.0 mass % (see Fig.2).

The possible synergetic effect indicated between P and Cu for TTS depends on Ni content. Dependencies of TTS of RPVM with high Ni content on Cu and P contents are shown in Fig.3. For RPVM with low Ni content, the dependencies of TTS on P and Cu content have other behaviour (see Fig.4).

The processes RA are complex and various RPVM exhibit specific behaviour of RA.

For RPVM of typical Western manufacture, the basic process of RA considers the formation and growth of Cu-enriched precipitates (Odette & Lucas 1986, Steel 1985, Hawthorne 1983).

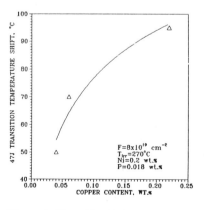

Figure 1. Transition temperature shift of RPVM WWER-type as a function of copper content

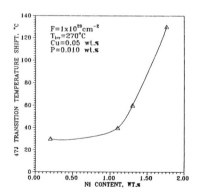

Figure 2. Dependence of the transition temperature shift of RPVM WWER-type on nickel content

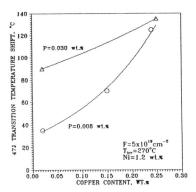

Figure 3. Dependence of the transition temperature shift of RPVM WWER-type with high Ni content on copper and phosphorus contents

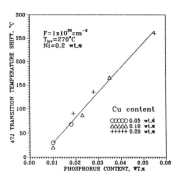

Figure 4. Dependence of transition temperature shift of RPVM WWER-type with low Ni content on copper and phosphorus content

In particular, in Guide 1.99 Revision 1 of the US NRC, the transition temperature shift (TTS) in degrees Celsius is represented by the equation:

$$\Delta T_F = \frac{5}{9}\{40 + 1000(Cu - 0.08) + 500(P - 0.008)\}\left(\frac{F}{10^9}\right)^{\frac{1}{2}} \tag{1}$$

where the values of Cu and P are in mass % and F is the fast neutron fluence (E> 1MeV). In an analysis of US surveillance data base, the general form of the TTS model depends on the effective impurity content (Odette & Lucas 1986), which is given by:

$$I_c = Cu\{0.6 + 0.4[1 - exp(-0.55Ni)]\} + 3P - 0.125 \tag{2}$$

where the values of Cu, Ni and P are in mass %.

In France, for example, each embrittlement result (transition temperature shift, TTS) obtained by the surveillance program is compared to the damage estimate based on an empirical formula, developed by Framatome for products used in France (van Walle, Fabry, Puzzolaute, Pouleure, Verstrepen & Van de Velde 1993):

$$\Delta T_F = 8 + \{24 + 1537(P - 0.008) + 238(Cu - 0.08) + 191Ni^2Cu\}\left(\frac{F}{10^{19}}\right)^{0.35} \tag{3}$$

where the values of P, Cu and Ni are in mass % and F = the fast neutron fluence, of energy (E) E> 1MeV.

An analysis of Russian material data base on surveillance samples gave relationships which indicated synergetic effects between the P and Cu influence on radiation embrittlement (Amayev, Kryukov, Levit & Sokolov 1993). For $\phi = 2.7 \cdot 10^{12}$ /cm²s (E>1 MeV):

$$\Delta T_F = 4.7(\phi t)^{0.32} + 8Cu(\phi t)^{0.32} + 250P(\phi t)^{0.38} + 2700PCU(\phi t)^{0.31} \tag{4}$$

where P and Cu are given in mass %.

All relationships have two aspects of differences in behaviour of RA of irradiated materials. First of all, there are differences in mechanisms of impurity precipitates formation at

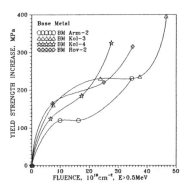

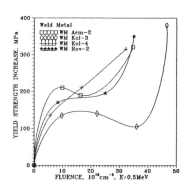

Figure 5a. Yield Strength increase of different materials (base metal only) from 4 NPPs of WWER-type as a function of neutron fluence.

Figure 5b. Yield strength increase of weld metal from 4 NPPs of WWER-type as a function of neutron fluence.

irradiation and, as a consequence, differences in radiation hardening. Secondly, there are differences in connection of radiation hardening with radiation embrittlement.

It is important to note that a main mechanism of radiation ageing can be the same for all kinds of RPVM. One consequence of this can be a result (Levit, Korolev, Tipping & Lessa 1997), which shows the same behaviour of radiation hardening for different materials and which have normal behaviour (with saturation and even softening) peculiar to thermal ageing. Fig. 5 a-c show radiation strengthening of different RPVM.

Radiation hardening of the metal under irradiation occurs primarily due to the formation of small-sized impurity precipitates. It can be, for example, small-sized copper precipitates or phosphides etc.. In recent and numerous microstructural researches, with the help of APFIM

(Atom Probe Field Ion Microscopy) and SANS (Small Angle Neutron Scattering) (van Walle, Fabry, Puzzolaute, Pouleure, Verstrepen & Van de Velde 1993, Miller & Burke 1990, Hawthorne 1991, Buswell, Little, Jones & Sinclair 1986) the presence of many types of different features was shown.

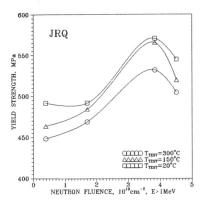

Fig. 5c. Effect of testing temperature on the yield strength as a function of fast neutron fluence: Reference base material JRQ (Final Database Report on IAEA CRP, 1994)

These were, for example, roughly spherical and disc -shaped precipitates rich in Cu and clusters which contained Ni, Mn, Si and P; Cu atmospheres; Cu phosphides; P clusters; spherical and rod-shaped Mo-carbides; Mo-phosphides; disk-shaped Mo-nitrides; Fe-nitride precipitates; P, Cu, Ni and Si precipitates located in the vicinity of the grain boundaries and in

the grain bulk. Phosphorus atmospheres have also been detected in the grain bulk, on the grain boundaries and also on the boundaries of coarse Mo carbides; P clusters, enriched with Ni and C atoms have also been detected. As these precipitates grow in size, a decrease in the bulk concentration of the impurity
levels dissolved in solid solution occurs. Smaller precipitates dissolve at the expense of larger ones and eventually the size distribution is such that the maximum effect (precipitation hardening) is present.

3 ANNEALING BEHAVIOUR

An important factor in the recovery of the irradiated RPVM properties is the annealing temperature. Results which were obtained using weld metal specimens containing 0.023% P and 0.12% Cu and irradiated at a NPP to a fast (energy, E > 0.5 MeV) neutron fluence (F) of $1 \cdot 10^{20}$ /cm^2 at 270°C and annealed at different temperatures during 150 h are presented in Fig. 6. The isochronal experiments indicate that the degree of DBTT recovery becomes larger as the annealing temperature (T_{ann}) increases relative to the irradiation temperature.

It should be noted that the most important parameter for the degree of DBTT recovery is the difference between the annealing (T_{ann}) and irradiation (T_{irr}) temperatures (see Fig.7). As can be seen, the difference ($T_{ann} - T_{irr}$) of 150 - 200 °C can give 50 - 100 % relative degree of DBTT recovery. This dependence, however, cannot be universal, because it must depend on the specific defect structure generated in the specific material (sensitivity to irradiation); different types and quantities of material-specific defects will require different activation energies to anneal out. This will cause different responses to annealing treatments applied.

Results of postirradiation annealing experiments for different RPVM of WWER-type are presented in Fig.8. The irradiation temperature was 270 °C for all of these materials and the annealing temperature was 420-460 °C .

As can be seen, the residual transition temperature shift (RTTS) does not depend on neutron fluence and does not exceed 50 °C ; for most materials it is less then 20 °C.

Previous work on annealing has shown, that the processes, occurring at the recovery stage of annealing, essentially depend on the annealing temperature (T_{ann}) (Amayev, Kryukov, Levit, Platonov & Sokolov 1996). At relatively low annealing temperature (i.e. 340 °C), insignificant recovery of irradiated RPVM properties occurs, but subsequent embrittlement at repeated irradiation (reirradiation) appears insignificant. On the contrary, at   a   higher relative annealing temperature (420- 460°C), more recovery of the properties is observed, but also

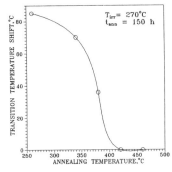

Fig.6 Transition temperature shift recovery of irradiated weld metal vs. annealing temperature

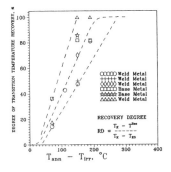

Fig.7 Dependence of the transition temperature recovery degree on the differences between T_{ann} and T_{irr} .

211

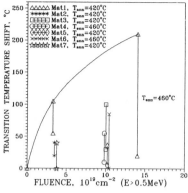

Fig. 8. Residual transition temperature shift of irradiated reactor pressure vessel materials after annealing at 420-460°C

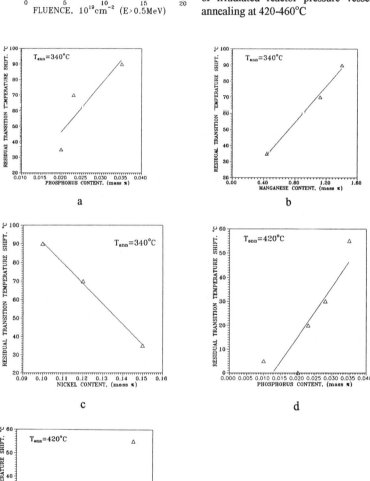

Fig.9 (a-e). Dependence of the residual transition temperature shift (RTTS) on the different element content of the irradiated reactor pressure vessel materials after annealing.

RE at re-irradiation condition occurs relatively rapidly. This indicates that essentially different processes are occurring at various temperatures. As most recovery processes are thermally activated, an increase in temperature is likely to activate additional processes of recovery and embrittlement and these new mechanisms should be taken into the overall assessment of RPVM annealing. As a result, it is expected that the explanation for the various dependencies of properties recovery by annealing will be found in the chemical composition of the respective alloys under consideration. At T_{ann} =340 °C, direct dependencies of RTTS on the P, Cu, Mn and Si contents and inversely direct dependencies on the Ni, Cr and Mo contents are observed (examples of this results are shown on Fig. 9 a-c). At high-temperature annealing (T_{ann} > 420 °C), the direct dependence of RTTS on the P contents (Fig. 9 d) and weakly indicated direct dependence on the Mn contents are observed (Fig. 9e).

The dependence on other elements is not revealed. The various influence of the contents of impurity levels on RTTS indicates that there are variety of processes occurring at different annealing temperatures.

The observed phosphorus effect on the residual TTS can occur for two main reasons. First of all, hardening of the bulk grain structure as a result of point defects accumulation, formation of dislocation loops, microvoids, various impurities precipitates, carbide and phosphide formation etc. can occur due to irradiation. Additionally, embrittlement of grain boundaries (decohesion) can occur due to segregation there of such impurities as P. The role of nickel is not clear, but it may be contributing to grain boundary embrittlement via nickel phosphide formation at the grain boundaries. Numerous research on steels with relatively high copper (> 0.1 weight %) and nickel (> 0.6 weight %) contents have demonstrated that one of the important hardening and embrittling processes occurring in such irradiated steel is the formation of copper-rich precipitates (Odette & Lucas 1986, Steel 1985, Hawthorne 1983, Tipping, Weber & Mercier 1987).

Research on grain boundary element structure was conducted by the method of a secondary ion mass spectrometry (SIMS) for RPVM before and after irradiation (Tursunov, Platonov & Levit 1987). Fig.10 presents the dependence of the relative intensities (I_P/I_{max}) of phosphorus from removed layers. This Fig.10 indicates, that while in an unirradiated material the phosphorus distribution is fairly uniform near to the grain boundaries, in a steel irradiated at T_{irr}=260°C up to fluence F=1.2·10^{20} /cm^2 (E>0.5 MeV), a pronounced phosphorus segregation is detected on the grain boundaries. Supported by the above observation, for recovery of the mechanical properties by annealing it is necessary, therefore, to create such conditions to not only dissolve the intergranular and grain boundary segregation but to heal out point defects and to overage precipitates (copper-rich) as well. Theoretically, for the complete recovery of a structure and properties of a material, realization of all annealing processes is required.

Special investigations on irradiation and annealing of more than 10 various RPVM, including

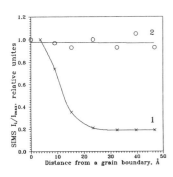

Fig. 10. Phosphorus distribution near grain boundary in the irradiated RPVM WWER-type (1 - F=1.2x10^{20} cm^{-2}, E>0.5 MeV; 2 - unirradiated material)

213

materials with the high contents of a phosphorus was conducted to get the phosphorus content effect on ΔT_{res}. The contents of a phosphorus changed within the limits of 0.010 - 0.0525 mass.%. The contents of a copper was within the limits of 0.03-0.26 mass. %.

The materials were irradiated by fluence of fast neutrons (E> 0.5 MeV) about 8×10^{19} cm^{-2}. Annealing of irradiated specimens was conducted at temperatures 460 - 470 C during 72-100 h. Linear regression analysis of results has given following empirical dependence:

$$\Delta T_{res} = -7 + 750 \cdot P \pm 10 \quad , \; C \tag{5}$$

where P is the phosphorus contents in a weight. %.

It is indicated that the residual TTS is associated with grain boundary phosphorus segregation formed during irradiation and subsequent annealing.

Microstructural investigations of irradiated and annealed RPVM (Hawthorne 1991) have shown, that in the process of annealing, a decrease in matrix concentration and increase of the size of the copper-rich precipitates occurs. Also, small-sized phosphides and carbides evolve within the grains (bulk effects). Annealing causes a softening and hence, a relative increase in the viscosity of the bulk grain body. On the other hand, at present there is a scarceness of data concerning the disappearance of grain boundary phosphorus segregation formed at irradiation. Thermodynamically, the dissolution and coagulation of small-sized segregation and defects are favoured with an increase in (annealing) temperature. This general behaviour has been confirmed by experimental data which shows that ΔT_{res} is reduced with increasing annealing temperature.

A relatively high enough annealing temperature (depending on irradiation temperature and alloy impurities) and enough time (kinetics of diffusion) will cause the precipitates to overage (grow) and a bulk grain softening can be expected. When embrittlement effects remain after annealing then they must be associated with grain boundary phosphorus enrichment (decohesion) effects. More research is required to resolve this aspect.

4 RE-IRRADIATION

For practical annealing applications to regenerate RPVM mechanical property levels to remain within the allowed limits, it is extremely important to establish the possible dependencies of radiation embrittlement of RPVM at reirradiation.

Figures 11-12 show empirical relationships of the transition temperature shifts of different RPVM on neutron fluences during re-irradiation after annealing at 420°C , 150 h.

In a previous work (Amayev, Kryukov, Levit, Platonov & Sokolov 1996) was shown, that the re-embrittlement behaviour could be described by vertical, lateral or conservative shift of the initial embrittlement dependence. The most restrictive (conservative) shift approach is used in the assessment of residual lifetime of annealed RPV. Experimental data have shown, that the so-called lateral shift approach could be used to remove unnecessary conservatism (Fig. 13)

As can be seen on Fig. 11-12 b, logarithmic analyses give the same behaviour of initial and post-annealing dependencies for saturation stage.

5 CONTROL METHODS FOR THE EFFICIENCY OF ANNEALING

A partial recovery of the mechanical properties of RPVM occurs during the process of annealing. The measurement of RPVM mechanical properties before and after annealing is necessary for the control of the RPVM condition after annealing and forecasting of RPV prolonged operation time. One method for this purpose, is to use a special device which enables

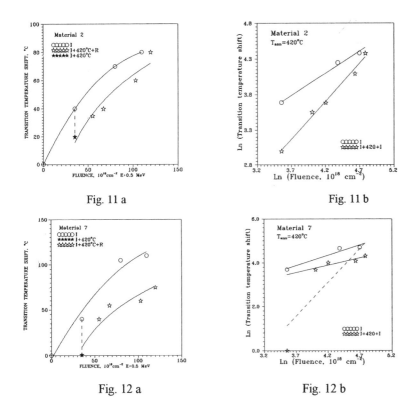

Fig. 11 a

Fig. 11 b

Fig. 12 a

Fig. 12 b

Fig . 11 a, b and 12 a, b represent dependencies of transition temperature
shift of various RPVM after annealing at reirradiation on neutrons fluences
in a normal and logarithmic scale.

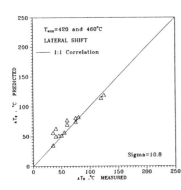

Fig.13   Comparison of measured
Charpy transition temperature shift
after re-irradiation with that predicted
by lateral shift method.

trepans to be cut out of the vessel inner surface. The trepan sample sizes permit the
determination of the chemical composition and the fabrication of subsized Charpy specimens
for mechanical tests. Subsized Charpy specimens tests may exhibit a large scatter and require
the establishment of an empirical correlation with standard Charpy specimens tests results.
This may be problematic. The trepan samples may also potentially reduce the strength of the
RPV and may create stress concentrators on the vessel . inner surface. Due to these

considerations, it is indicated that a non-destructive hardness test method is used on the vessel wall before and after annealing, instead of taking physical specimens away from it. When hardness measurements on the vessel inner surface are impossible, for example, when the vessel inner surface is cladded with stainless steel it is still possible to perform the hardness measurement in the vessel outer surface.

Measurements conducted before and after annealing have shown a good correlation of hardness measurement results with the destructive methods results and have shown that the recovery of weld metal according to the hardness change is more marked comparative to that of the base metal. Development of hardness measurement methods, as well as other physical non–destructive methods (such as methods of RPVM magnetic properties, conductivity and other measurement), which enable evaluation of change of physical and mechanical properties important for RPVM safe operation, favour a more correct assessment of RPVM life during normal operation and after annealing.

6 CONCLUSIONS

1. There are many mechanisms responsible for RPVM radiation ageing (RA) under fast neutron irradiation at about 270 - 300 C. This has lead to a variety of empirical laws being used in various countries to assess it.
2. RA of all type of RPVM have the same principal mechanism: impurities redistribution, formation and growth of impurities precipitates. And therefore RA regularities are typical for all ageing processes. But due to different chemical composition all materials have there
own peculiarities of RA.
3. Optimum annealing parameters (time at temperature) have the potential to cause partial recovery of some RPVM mechanical properties which have degraded due to RA.
4. The model for RA processes for repeated irradiation after annealing which appears to best describe the observations is the so-called lateral shift one.
5. The further improvement and development of non-destructive methods for physical and mechanical properties measurement is necessary. In particular, hardness methods need more correlation and benchmarking activities to enhance their use in safety assessments of RPVM operation after annealing.

REFERENCES

Amayev A.D., Kryukov A.M., Levit V.I., Sokolov M.A. 1993. Radiation stability of VVER-440 vessel materials. "Radiation Embrittlement of Nuclear Reactor Pressure Vessel Steels: An International Review (Fourth Volume), *ASTM STP 1170*, Lendell E Steel, Ed., American Society for Testing Materials, Philadelphia.

Amayev A.D., Kryukov A.M., Levit V.I., Platonov P.A., Rogov M.F. and Sokolov M.A. 1993. Radiation Damage and Recovery in VVER-440 Vessels Materials. *Proceedings of PLEX-93*. Zurich.

Amayev A.D., Kryukov A.M., Levit V.I., Platonov P.A. and Sokolov M.A. 1996. Mitigation of irradiation embrittlement by annealing. Effect of radiation on materials: *17th International Symposium, ASTM STP 1270*, David S.Gelles, Randy K. Nanstad, Arvind S. Kumar and Edward A. Little, Eds. American society for Testing and Materials, pp 232-247.

Beaven P.A., Frisius F., Kampman R and Wagner R. 1986. *Proceedings of the 2nd International Symposium on Environmental Degradation of Materials in Nuclear Power*

Systems - Water Reactors. (American Nuclear Society, La Grange Park, IL), p. 396.

Brauer G. and Eichhorn F. 1993. Considerations about irradiation-induced precipitates in Soviet type reactor pressure vessel steels. *Nuclear Engineering and Design*, 143. pp 301-307.

Buswell J.T., Little E.A., Jones R.N. and Sinclair R.N. 1986. *Proceedings of the 2nd International Symposium on Environmental Degradation of Materials in Nuclear Power Systems - Water Reactors.* (American Nuclear Society, La Grange Park, IL). p. 139.

Final Database Report on IAEA CRP on Optimising of Reactor Pressure Vessel Surveillance Programmes and their Analysis. 1994. Volume II, Vienna, Austria.

Hawthorne J.R. and Steele L.E. 1976, In: Effects of Radiation on Structural Metals, *ASTM STP 426.* (American Society for Testing and Materials, Philadelphia), p. 534.

Hawthorne J.R. 1983. In: Embrittlement of engineering alloys. Treatise on Materials Science and Technology, eds. C.L.Briant and S.K. Banerji. v. 25 (New York, London) p. 423.

Hawthorne J.R. 1991. SANS Investigations of RPV Steels and Welds in Irradiated, Annealed and Reirradiated conditions. JCC CNRS, USA, October.

Korolev Ju. and Levit V. 1993. Radiation Embrittlement and Hardening of Materials of VVER-440 Vessels. *Proceedings of Joint Seminar Russia and USA NRC WG-3-10/93.* USA.

Levit V.I. Yu.N. Korloev, Ph. Tipping and R.N.T. Lessa. 1997 Empirical correlation of observed three stages of fast neutron irradiation hardening and embrittlement in WWER-440 pressure vessel materials. Effects of radiation on materials: 18th International Symposium *ASTM STP 1325*, Randy K Nanstad, Ed. American Society for Testing and Materials, (to be published).

Miller M.K. Hoelzer D.T., Ebrahimi F, Hawthorne J.R. and Burke M.G. 1987. *J de Physique.* 48-C6, p.423.

Miller M.K., Hoelzer D.T., Ebrahimi F., Hawthorne J.R. and Burke M.G. 1988. *Proceedings of the 3rd International Symposium on Environmental Degradation of Materials in Nuclear Power Systems - Water Reactors.* eds. G.J. Trends and J.R. Weeks (The Metallurgical Society Inc., Warrendaly PA). p. 133.

Miller M.K. and Burke M.G. 1990. *Proceedings of the 14th International Symposium on the Effects of Radiation on Materials.* eds. N.H.Packan, R.E. Stroller and A.S. Kumar. *ASTM STP 1046*, v.2, (American Society for Testing and Materials, Philadelphia) p. 107.

Miller M.K. Jayaram R. Other P.J. and Brauer G. 1994. APFIM Characterisation of 15Kh2MFA Cr-Mo-V and 15Kh2NMFA Ni-Cr-Mo-V type steels. *Applied Surface Science,* 76-77, pp 242-247.

Nanstad R.K., Iskander S.K., Rowcliffe A.F. Corwin, W.R. Odette G.R. 1990. *Proceedings of the 14th International Symposium on the Effects of Radiation on Materials.* eds. N H Packan, R E Stroller and A S Kumar. *ASTM STP 1046*, Philadelphia. p.5.

Odette G.R. and Lucas. G.E. 1986. Irradiation Embrittlement of Reactor Pressure Vessel Steels: Mechanisms Models and Data Correlations. *ASTM STP 909*, Philadelphia, pp 206-241.

Serpan C.Z. 1973. In: Effects of Radiation Substructure and Mechanical Properties of Metals and Alloys. *ASTM STP 529*, p.92.

Steele L.E., Davies L.M., Ingham T. and Brumovsky M. 1985. In: *Effects of Radiation on Materials. ASTM STP 870* (American Society for Testing and Materials, Philadelphia), p 863.

Tipping Ph, Weber W. and Mercier O. 1987. A study of the mechanical property changes of irradiation embrittled pressure vessel steels and their response to annealing treatments. *Trans. 9th International Conf. on Structural mechanics in Reactor Technology (SMiRT).* e.d. F Wittman, Luasanne. vol. G. pp 115-127.

Tipping Ph., and Weber W. 1989. Irradiation and annealing effects on the mechanical properties of a RPV steel and associated PLEX considerations. *Trans. 10th Internat. Conf. on Structural mechanics in Reactor Technology (SMiRT);* ed. A. Hadjian, Los Angeles, vol G. pp 239-244.

Tursunov I. Ye, Platonov P.A. Levit V.I. 1987. Role of radiation stimulated process in the steel embrittlement. *IAE-4478/11.* Moscow, 37 pp.

Walle van E. Fabry A. Puzzolaute J-L, Pouleure Y. Verstrepen A. Van de Velde J-P. 1993. Belgian contribution to the IAEA Phase 3 Coordinated Research Programme on "Optimisation of Reactor Pressure Vessel Surveillance Programmes and Their Analyses". IAEA, Vienna

ACKNOWLEDGEMENT

The valuable discussions with Dr. Ph. Tipping concerning all aspects of this report are gratefully acknowledged.

DISCLAIMER

The ideas and views expressed herein by the author are his own and these do not necessarily have to agree with those of his affiliation.

Ageing of Materials and Methods for the Assessment of Lifetimes of Engineering Plant, Penny (ed.)
© 1997 Balkema, Rotterdam, ISBN 90 5410 874 6

Assessment of operation-dependent structural life of ageing aircraft using crack growth retardation model

Jae-Young Jeon
TeraSource Venture Capital Co. Ltd, Korea

Jyung-Yil Lee
Sogang University, Korea

ABSTRACT: There are many aircraft being operated by near to or over the design lifetime due to various economical circumstances. A kind of reducing mission load strength for the weakened structure can be considered as a means of the structural integrity management of the ageing fleet. In this study, the damage tolerance of a structure with crowing crack is assessed by considering the effect of reducing mission load. The analyzed results show that the peak load limited type operation can increase the crack growth rate, resulting in shortening of the remaining life.

1 INTRODUCTION

Aircraft life management activities include the design, maintenance and economic discard. In the Korean Air Force, about 60% of the force management cost, including the initial purchasing, is used for structural integrity support (Lee 1990). Aircraft structures are exposed to the continual fatigue loads, which are originated from the various load source depending on the operation mode. For the optimization between the structural safety and weight limitation, the damage tolerance concept has been widely used for aircraft structure design (Fuches & Stephens 1990). The damage tolerance of an aircraft structure requires that the main structure with damage (crack) can safely operate until the damage can be defected to repair. Any structure is assumed to have the initial damage smaller than the detectable size which depends on the inspection method and severity. For safe operation, the structural design life and maintenance interval is to be determined by comparison between the initial and critical crack size and crack growth rate (Kwon 1995), (Broek 1988), (ASIP 1975), (MIL-A83444 1974), (Airworthiness Standards 1994), (Jeon 1995).

The critical crack size is calculated by the fracture mechanical consideration as the crack size until which the structure can sustain the maximum load without instantaneous propagation of the crack. The crack growth rate should be calculated by considering the repeated load history and crack retardation effect. The crack retardation effect becomes dominant when the proceeding peak load level is considerably higher than the following loads. In this case, the plastic zone developed by the preceded big load hinder the crack opening at the following loads that it becomes difficult for the crack to grow. A big peak load has 2-face effect simultaneously; the one for crack propagation by the load itself and the other for the retardation of the crack growth by the following loads. As a result, the load reduction method for an aged aircraft structure operation should be carefully considered for adopting as a means of life extension because the weakened retardation effect due to the lessened peak load causes fast crack growth, i.e. early failure.

In this study, F-5 A/B, E/F of KAP which were manufactured during the 1970's shall be used for analyzing model. They are originally designed for 4,000 hrs (20 yrs) operation life. The analyzing point of the structure is defined as W.S. 39.5 at 44% Spar, Lower Flange Fastener Hole, which is one of the Fatigue Critical Locations in the main wing (Fig 1). The stress analysis result of the full scale Aircraft is from the original designer, Northrop. The load history is based on the stress spectrum derived from the load exceedence curves developed from the flight load data record (SWRI 1978).

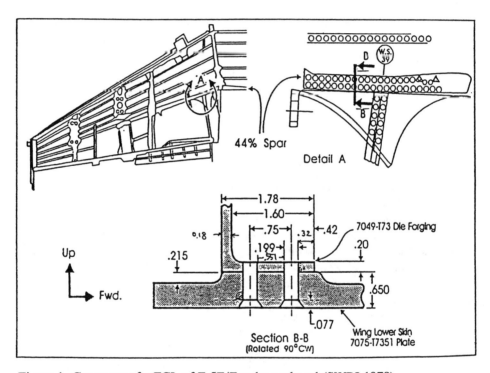

Figure 1. Geometry of a FCL of F-5E/F to be analyzed (SWRI 1978)

2 CRACK GROWTH LIFE DEPENDING ON OPERATION MODE

2.1 Determination of the Load Spectrum at a Fatigue Critical Location

The accumulated stress spectra at the FCL which were reduced from flight load records are shown in Fig. 2. After applying the truncation and clipping technique, the maximum stress amplitude and corresponding occurrence are calculated as Table 1.

220

The minimum stress is taken as zero for conservative analysis assuming load application sequence to be random. And, the used manipulating parameters are:
clipping stress   = 28 ksi
truncating stress   = 2 ksi

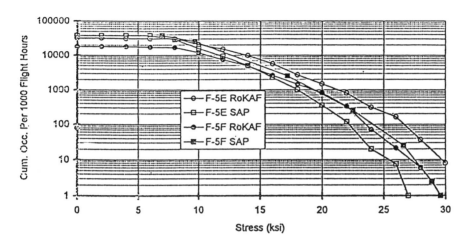

Figure 2. Stress spectra converted from flight data (SWRI 1978)

Table 1. Calculated fatigue stress and occurrence reduced from exceedence curve (for 1,000 flights ; S_{min} = 0

S_{max}(ksi)	2	4	6	8	10	12	14
cycles	31,771	31,227	30,711	30,193	20,888	14,984	9,779

S_{max}(ksi)	16	18	20	22	24	26	28
cycles	5,644	2,706	1,459	794	305	162	47

The reference stress at the analyzing point was calculated from the relation (SWRI 1978)

$$\sigma_{ref} = 0.0155M_x$$

Where M_x is the bending moment applied to the root rib at W.S. 39.5. In the relation above, the stress is linearly dependent on the applied bending moment, which is the reason why the analyzed FCL point is selected. It can be conceived that the bending moment is linearly dependent on the mission load strength too.

221

2.2 Load Reduction by Operation Mode Modification

In this study, three reduced load level of 90, 80, 70% is used with respect to the initial design load (100%). Also, two types of load reduction are considered; the one is for the case of total load reduction in which every mission strength is reduced by the reduced load level, the other is for peak load reduction in which only some loads exceeding the reduction ratio of the original clipping stress are reduced (as shown in Table 2).

Table 2. Analyzed Types of Applied Load Spectra

Operation Mode	normal operation	totally reduced	peak load limited
Type of Load Spectrum			
Operation Strategy	normal stress spectrum	all operations are under reduced loading	Only a specially heavy-loaded mission limited, like as hit and run- away
Remarks	for F-5E RoKAF	mission change	usually considered

2.3 Materials Characteristics and Initial Crack Size

Through the originally used material for the Wing Spar was extruded 7049-T3, the property of die forged material is used here because it presumable has similar properties to the original and often practically used for replacement (SWRI 1978), (MIL-HDBK-5G).

$$K_c = 59 \text{ ksi-in}^{\frac{1}{2}} \text{ for mixed mode}$$

$$\sigma_y = 60 \text{ ksi} \qquad \sigma_{uts} = 71 \text{ ksi}$$

According to the handbook (Gallagher et al 1984), the initial crack size is 0.05" for the slow crack growth structure without redundant load path. Also, the effect of cold working is considered using the shrinked initial crack size, $a_i = 0.005$". The used properties for fatigue crack growth are shown in Table 3.

Table 3. Used Material Property of Fatigue Crack Growth Rate (Al 7049-T73; DIE FORGING; LT; Lab. Air; 70F/21C)

ΔK	da/dN[in/cycle] for R[S_{min}/S_{max}]					
[ksi√in] R= 0.0	R= 0.1	R= 0.2	R= 0.3	R= 0.4	R= 0.5
1	3.600000E-09	4.172179E-09	4.920129E-09	5.931510E-09	7.360220E-09	9.500457E-09
3	1.683553E-07	1.951135E-07	2.300917E-07	2.773893E-07	3.442034E-07	4.442924E-07
5	1.006231E-06	1.166160E-06	1.375218E-06	1.657907E-06	2.057244E-06	2.655459E-06
8	5.213357E-06	6.041962E-06	7.125108E-06	8.589744E-06	1.065874E-05	1.375813E-05
10	1.13842E-05	1.319359E-05	1.555881E-05	1.875708E-05	2.327506E-05	3.004308E-05
12	2.154948E-05	2.497453E-05	2.945173E-05	3.550583E-05	4.405803E-05	5.686943E-05
20	1.287975E-04	1.492684E-04	1.760279E-04	2.122122E-04	2.633272E-04	3.398987E-04
24	2.438046E-04	2.825546E-04	3.332083E-04	4.017026E-04	4.984597E-04	6.434041E-04
27	3.681931E-04	4.267133E-04	5.032105E-04	6.066503E-04	7.527728E-04	9.716675E-04
30	5.323863E-04	6.170031E-04	7.276137E-04	8.771819E-04	1.088467E-03	1.404976E-03
55	4.441932E-03	5.147927E-03	6.070800E-03	7.318712E-03	9.081554E-03	1.172233E-02
60	6.023264E-03	6.980594E-03	8.232011E-03	9.924180E-03	1.231460E-02	1.589549E-02
65	7.970751E-03	9.237613E-03	1.089365E-02	1.313294E-02	1.629624E-02	2.103494E-02
70	1.033108E-02	1.197309E-02	1.411951E-02	1.702191E-02	2.112195E-02	2.726388E-02
75	1.315276E-02	1.524324E-02	1.797591E-02	2.167104E-02	2.689089E-02	3.471034E-02
80	1.648608E-02	1.910636E-02	2.253157E-02	2.716316E-02	3.370588E-02	4.350703E-02

2.4 Stress Intensity Factor and Geometry Factor

The applied stress intensity factor is calculated based on the reference stress value at the analyzing point, where the shape factor, β, is obtained by simplifying the rivet hole as an opening in the finite plate as

$$K_I = \beta_{total} \sigma_{ref} \sqrt{\pi a}$$

The actually applied stress is composed of the tensile stress normal to the crack plane and the transmitted load by the rivet to the plate. Using the previous study about the rivet hole analysis (SWRI 1978)

$$\sigma_{bear} = 0.471\sigma_{ref}$$

As a result, the total geometry factor is calculates as, shown in Fig. 3

$$\beta_{total} = 0.471\beta_{bear} + (1+\alpha)\beta_{grad}\beta_{tens}$$

where $\alpha = 0$ (for $a>t$) and 0.03 (for $a<t$)

$\beta_{grad} = 1.039 - 0.0194\ a/t$ (for $a<t$) 1.02 (for $a \geq t$)

223

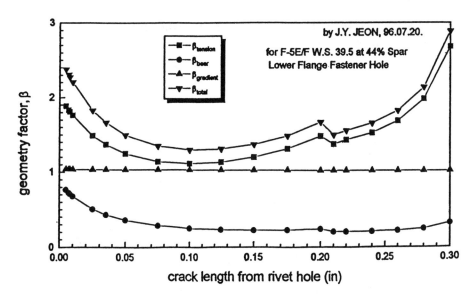

Figure 3 Geometry Factor Calculated

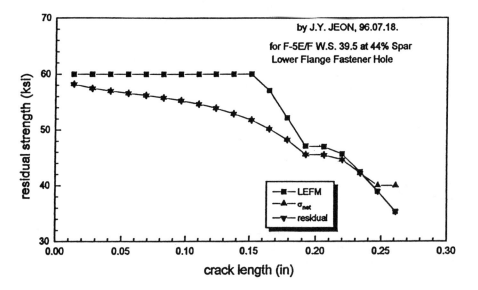

Figure 4. Residual Strength

2.5 Residual Strength

The stress value at the analyzing point when the crack instantaneously propagate is calculated from the fracture toughness of the material and the geometry using the linear elastic fracture mechanics as;

$$\sigma_{res} \langle \frac{K_c}{\beta_{total} \sqrt{\pi a}}$$

224

Also, applying the constriction that the average stress at the remained segment of the cracked plate can not be greater than the material yield strength;

$$\sigma_{net} = \sigma_{res} \frac{W-a}{W} \langle \sigma_y$$

The residual strength is obtained by combining two equations above, as shown in Fig. 4.

3 RESULTS

3.1 Growth Life Neglecting Retardation Effect

The crack growth results are shown in Fig 5. for the normal load condition (100%) and the reduced cases without considering the retardation effect. the critical crack size is taken as 0.25" irrespective of the load condition, which corresponds to the 80% of the plate ligament. Even though the structure can bear the crack size larger than this, it can not continue to operate due to the maintenance custom. Actually, the time duration after this size has little meaning because the crack propagation occurs so quickly.

In Fig. 5, the time required for the crack to grow to failure by the peak load reduced operation is nearly the same to the normal operation. This is because the number of load cycles whose strength should be reduced by peak load limitation are not so many. On the other hand, the life is prolonged for the case of total load reduction case. It is known hat the crack growth life is mainly depend on the small load repetitions with many occurrences rather the big load with little repetitions when the retardation effect is not considered.

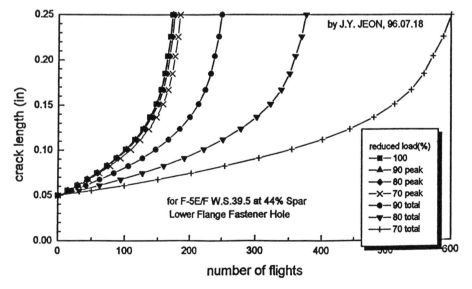

Figure 5. Crack Growth without Retarded Effect

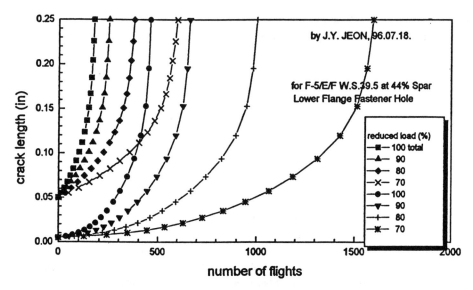

Figure 6. Initial Crack Size Effect

The crack growth results for the initial crack sizes of a_i = 0.05" and 0.005" respectively, are shown in Fig. 6. The crack growth life for a_i = 0.005" increases twice more than that of 0.05". This means that the quality control or the inspection technique which can shrink the initial crack size as small as possible is important for the structural safety. These are related to NDT, shape design, process design, etc.

3.2 Crack Growth Life Considering the Retardation Effect

The crack retardation effect is considered by Willenborg model without correction factor (Willenborg et al 1971), (Forman et al 1957). The material properties for crack growth in Table 3 are used with the modified stress intensity factor as:

$$K_{reduced} = K_{max.req} - K_{max.i}$$

$$K_{maax.eff+} = K_{max} - K_{reduced}$$

$$K_{min.eff} = K_{min} - K_{reduced}$$

$$R_{eff} = K_{min.eff} / K_{max.eff}$$

The crack growth results are shown in Fig. 7. The crack growth life increases for the case of the total load reduction, on the other hand it decreases for peak load reduction case. The amount of the life extension or reduction becomes greater proportionally to the reduction ratio depending on the reduction types, i.e. total or peak, respectively.

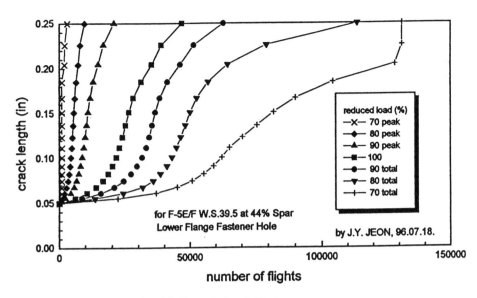

Figure 7. Crack Growth with Retardation Effect

Table 4. Crack Growth Life Willenborg Retardation Model

reduced peak load (%)	total load spectra reduced	peak load limited
100	46,721 (1.00)	
90	62,808 (1.34)	20,582 (0.44)
80	113,716 (2.43)	9,640 (0.21)
70	131,072 (2.81)	3,000 (0.06)

In Table 4, the structural life can be reduced to as minimum as 6% of the normal operation when the aircraft operation continue with 70% peak load reduction mode. The reason why the crack growth rate becomes high when the peak load is limited is that the crack retardation effect fades out.

4 CONCLUSION

By considering the crack growth life with retarded crack growth model, it was known that the conventionally used peak-load-reduced operation can decrease the remaining life. However, the maintainability becomes better because of the easy detectability of the grown crack before failure as the critical crack size is larger depending on the reduced peak load. This means that the peak load limited operation can be considered good only for easy accessible and frequently inspectable structure.

REFERENCES

Airplane Damage Tolerance Requirements, 1974. MIL-AA83444.
Airworthiness Standards 1994: Transport Category Airplanes, Code of Federal Regulations, Part 25, FAA.

ASIP, 1975. Airplane Requirements, MIL-STD-1530.

Broek, D. 1988. *The Practical Use of Fracture Mechanics*. Kluwer Academic Publishers, M.A.

F-5. 1978. Durability and Damage Tolerance Report, SWRI.

Foreman, R.G., Kearney, V.E. and Engle, R.M. 1957. (Numerical Analysis of Crack Propagation in Cyclic-Loaded Structure, *ASME Trans. J. Basic Eng.* vol. 89D. p. 459.

Fuches, H. O. and Stephens, R.I. 1990. *Metal Fatigue in Engineering*. John Wiley & Sons, NY.

Gallagher, J.P., Giessler, F.J. and Berens, A.P. 1984. *USAF Damage Tolerant Design Handbook*. AFWAL-TR-82-3073, Wright Patterson Air Force Base, Ohio, pp. 2.4.13.

Jeon, J.Y. 1995. "The Life Assessment of Engine Ma Major Components". Proc. *'95 Aeronautical Technology Seminar,* Korean Air Force, pp 35-48, Korea.

Kwon, J.H. 1995. "The Structural Integrity Management of an Ageing Aircraft for Operation Life Extenstion", Proc. *'95 Aeronautical Technology Seminar*, Korean Air Force, p. 49. Korea.

Lee, H.H. "Aeronautical Technology and Air Force". Proc. *1st Symposium on Aircraft Development*, pp 23-32, Agency for Defence Development, Korea.

MIL-HDBK-5G. pp 3-283.

Willenbong, J., Engle, R.M. and Wood, H.A. 1971. "A Crack Growth Retardation Model Using an Effective Stress Concept". AFDL-TM-71-1-FBR.

Ageing of Materials and Methods for the Assessment of Lifetimes of Engineering Plant, Penny (ed.)
© *1997 Balkema, Rotterdam, ISBN 90 5410 874 6*

Physical ageing causes of alloy metals in the vibration conditions and method for the phenomenon assessment

A.Jakowluk
Bialystok Technical University, Poland

ABSTRACT: For the design of the constructions which in the dynamical creep work conditions, the knowledge of a curve of the equal limiting dynamical creep strain (equal strain for the estabilished limiting time) in the dependence from σ_m - vibration mean stress and σ_a - stress amplitude is wanted. A problem of the phenomenon assesment is very complicated because the investigation results are dependent from of the: 1. material type, 2. temperature T, 3. cycle numbers N_f, 4. stress states. For the interpretation of the strain hardening or weakening of sample material the micropolar waves are applied. These waves stand up during the vibration action. For the fatigue in the complex stress states the Sdobyrev's strength criterion is proposed.

1 INTRODUCTION

Examining the fatigue strength of the different alloy steels at the constant temperature frequently qualitatively the mutable strength characteristics are affirmed or investigating. The same steel alloy but at the different temperatures qualitatively similar phenomena as in the first case are affirmed. Till now qualitatively these problems were interpreted applying the metal physics methods (mainly for the static loadings). Instead the superiority of the phenomena interpretation by the micropolar wave use gives the inductorly comparison of the motion equations, namely:
- classical mechanics.

$$\mu\nabla^2 u_i + (\lambda+\mu)u_{k,ki} + \rho b_i = \rho\ddot{u}_i \tag{1.1}$$

- micropolar mechanics

$$(\lambda+\mu)u_{j,ji} + (\lambda+\kappa)u_{i,ji} + \kappa_{e_{ijk}}\phi_{k,j} + \rho b_i = \rho\ddot{u}_i$$

$$(\alpha+\beta)\phi_{j,ji} + \gamma\phi_{i,jj} + \kappa_{e_{ijk}} - 2\kappa\phi_i + \rho M_i = j\ddot{\phi}_i \tag{1.2}$$

where: u_i - displacement co-ordinates; ϕ_i - microrotation vector co-ordinates; e_{ijk} - Levy-Civita tensor co-ordinates; M - volumeric moment co-ordinates; λ,M - Lame constants; $\alpha,\beta,\gamma,\kappa$ - additional constants in the miocropolar elastic body; $j_{kl} = j\delta_{kl}$ - co-ordinates of the rotational inertia tensor for the microisotropic body; ρ and j - constants in linear theory.

The simplified model of the mocripolar linear elastic body (Eringen 1968) establish the classical elastic materials which have the additional degree of freedom and which results only from the local rotation of the rigid particles.

For the investigations the uniaxial forced sample in the classical theory can obtain only two waves but for the micropolar theory - four wave (Parfit & Eringen (1966). In this paper is shown that the strain hardening or weakening of the different metal alloys are dependent from the micropolar waves.

2 MATERIAL TYPE INFLUENCE ON THE FATIGUE STRENGTH IN THE DYNAMICAL CREEP PROCESS

2.1 Harmonic loading for the uniaxial tension

The wibrocreep and fatigue material investigations regard of the cyclic loadings, i.e.

$$\sigma(t) = \sigma_m + \sigma_a \sin \omega t = \sigma_m(1 + A_\sigma \sin \omega t), \quad A_\sigma = \sigma_a/\sigma_m \tag{2.1}$$

where: σ_m - mean stress, σ_a - stress amplitude, A_σ - stress amplitude coefficient, $\omega = 2\pi f$ - angular velocity, f - frequency.
   The sample response on the dynamic loading by Eqs. (2.1) has the following form

$$\varepsilon(t) = \varepsilon_m(t) + \varepsilon_a \sin(\omega t - \varphi) \tag{2.2}$$

where: $\varepsilon_m(t)$ - mean strain changed at the time; $\varepsilon_a(t)$ - amplitude changed at the time; φ - loss angle. The stress course by Eqs. (2.1)$_1$ and strain by Eq. (2.2) is presented in Fig.2.1. The processes which in the sample can appear, by suitablely estabilished values of the σ_m and σ_a, are defined as follows:

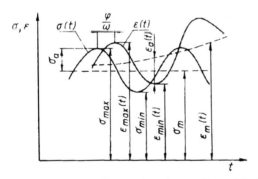

Figure 2.1. Graphical ilustration harmionic loading and displacement at the phase the material response

1) $A_\sigma = 0$ - a creep ($\sigma_m + 0$, $\sigma_a \neq 0$)
2) $0 < A_\sigma < A_\sigma^f$ - a vibrocreep
3) $A_\sigma^f < A_\sigma$ - a creep in the fatigue process for the no symmetrical cycle
4) $A_\sigma = \infty$ - a fatigue in the symmetrical cycle ($\sigma_m = 0$, $\sigma_a \neq 0$).
The ratio φ/w - phase displacement at the time the effect $\varepsilon(t)$ to cause $\sigma(t)$. The creep causes 2) and 3) produce the dynamic creep.
   For the design of the constructions which in the dynamical creep work conditions, the knowledge of a curve course of the equal limiting dynamical creep strain (equal strain for the estabilished limiting time) in the dependence from σ_m and σ_a is wanted. A problem of the phenomenon assessment is very complicated because the investigation results are dependend from of the: 1, material type, 2. temperature, 3. cycle number N, 4. stress state.
   Several authors, suming up the investigation results, two different curve types in the Haigh's system were offered (Fig. 2.2).
   On the Fig. 2.2 curve I, (Vitovec 1957) represents the metal alloy curves which by the dispersion are strengthened at the high temperatures but curve II, (Kennedy 1962) (Jakowluk 1967) represents the curves for the pure metals and simple alloys which at the lowered temperatures are weakened.

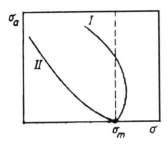

Figure 2.2. Possible curve types, characterized the σ_a and σ_m action, which to equal strain in creep for the establibished time or to identical life t_r leads: I - for metal alloys strain hardening at the high temperatures, II - for pure metals and simple alloys at the low temperatures.

2.2 Destructive action the vibration of small amplitude on the life of the simple aluminium alloy (AlMg5)

The main investigation aim is a demonstration that no a value of the stress amplitude σ_a decides of very great a change life of the AlMg5 alloy in the ratio to static creep but only the vibration existence.

The creep and vibrocreep tests on the flat samples, cuted out in the rolling direction by (6) Jakowluk. The test conditions were following: 1) $A_\sigma = \sigma_a/\sigma_m = 0.005$ for the different values σ_m, 2) $f = 31$ Hz, 3) $A_\sigma = 0$ for the different stress values $\sigma = \sigma_m$, 4) T - room temperature.

The research results are presentes on Fig. 2.3 on which are two creep, vibrocreep curve pairs, i.e.: "1-3", "2-4".

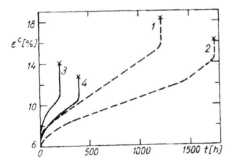

Figure 2.3. Creep (1,2) and vibrocreep (3,4) curves by $A_s = 0,005$ and $f = 31$ Hz; curve pair (3,1) - $s_m \approx s = 313$, 1 MPa; curve pair (4,2) - $s_m \approx s = 306,5$ MPa.

For the creep tests (1,2) and vibrocreep (3,4) the following conclusions results:
1) The application on the tension direction a very small vibrations ($\sigma_a = 0,005\sigma_m$) gave the considerable vibrocreep velocity increase (curves 3 and 4) in the ratio to static creep (curves 1 and 2) and simultaneous life t_r shortening from 4 to 6 times.
2) The failure fractures had completey different character for the creep and vibrocreep: a) when for the creep the fractures are smoth and under angle about 45° while for the vibrocreep were a multiplanes also under angles about 45° but with the tendency to the brittle fractures. From Fig. 2.3 results that vibrocreep curves 3 and 4 have the lower elongations for the rupture of 3 - 4%.

3 TEMPERATURE INFLUENCE ON THE MECHANICAL FATIGUE

The fundamental investigations on the fatigue temperature influence of the steel S - 816 alloy in the uniaxial tension by   Vitovec and Lazan were performed. The generalized Haigh's diagram

231

form in the σ_m/R_m - σ_a/σ_f co-ordinates was presented on Fig. 3.1. The curves show the curvature increase along with the increase temperature. In the lowest situation is the investigated curve at the room temperature, highest is situated the investigated curve at the 1173K. The experimental points obtained at the high temperatures lie above the circle of radius the $\sigma_m/R_m = 1$, nearly or above the curve.

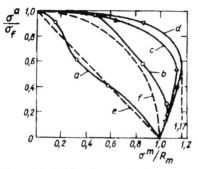

Figure 3.1. Haigh's diagrams in the nondimensional co-ordinates for the samples an alloy steel S-816 at the temperatures [K]: a -297, b - 1009, c - 1090, d - 1173, e - linear failure law, f - nonlinear law in the form a circle.

Presented investigation results of the alloy steel S - 816 for the different temperatures carry in the very inportant conclusions, namely:
1) The alloy steels at the elevated and high temperatures experience the considerable strain hardening at the vibration existence conditions.
2) For the economical regard the alloy steels it is necessary applay at the optimum work temperature.

4 WORK OF THE STRUCTURAL COMPONENTS IN THE COMPLEX STRESS STATES IN THE VIBRATION CONDITIONS AND HIGH TEMPERATURES

In the preface it is necessary mention that majority the structural components work in the complex stress states and are surrendered, if no immediate, this in direct the vibration interaction, instead the temperature is conditioned the technological processes.

4.1 Influence of the asymmetrical loading cycle on the behaviour of the different alloys under fatigue

For the harmonic tension and static torsion, i.e.in the biaxial stress states

$$\sigma_{11}(t) = \sigma_{11}^m + \sigma_{11}^a \sin \omega t, \quad \sigma_{12} = \text{const} \tag{4.1}$$

can note Eq. (4.1) in the form

$$\sigma_i(t) = \sigma_i^m(1 + A_{\sigma_i} \sin \omega t), \quad A_{\sigma_i} = \sigma_i^a/\sigma_i^m \tag{4.2}$$

where: σ_i^m - stress intensity og the mean stress tensor σ^m by the following definition

$$\sigma_i^m = (-3I_{2s}^m)^{1/2} = \left(\frac{3}{2}s_{ij}^m s_{ij}^m\right)^{1/2} \tag{4.3}$$

I_{2s}^m - second invariant of the stress deviator s^m, s_{ij}^m - deviator co-ordinates of the mean stresses, σ_i^a - intensity of the stress amplitude tensor σ^a.

For the vibrocreep description (Jakowluk & Mieleszko 1985) were leaded the substitute static stress tensor on the vibration direction and the modified stress intensity, i.e.

$$\sigma_{ij}^{s} = \sigma_{ij}^{m} + p\sigma_{ij}^{a}, \quad \sigma_{i}^{s} = \left[(\sigma_{11}^{m} + p\sigma_{11}^{a})^{2} + 3\sigma_{12}^{2} \right]^{1/2} \tag{4.4}$$

In this paper statistical was shown that for the metals for the small values of A_{σ_i} the mean value of parameter p one can accept 0.5.

In the dynamical creep processes in complex stress states how the correct strength criteriom for the metals one can assume the formulated criterion by (9) Sdobyrev's, i.e.

$$\sigma_{red} = \sigma_{1}\beta + (1 - \beta) \tag{4.5}$$

where: σ_{1} - maximal principal stress, $\sigma_{i} = f(I_{2s})$ - stress intensity, β - material constant for the isothermal and static processes of which the values are within a range of $0 \le \beta \le 1$, for $\beta = 1$ - britle rupture, $\beta = 0$ - ductile rupture. The generality of this criterion is due to the fact that it is function F depentent on the three invariants

$$F(I_{1\sigma}, I_{2s}, I_{3s}) = 0 \tag{4.6}$$

For the vibrocreep and fatigue conditions the Sdobyriev's criterion by Eq.(4.5) is modified. Namely, material constant β exceeds on the material function, i.e.

$$\beta = \beta(A_{\sigma_i}) \tag{4.7}$$

4.2 Vibration action effects in biaxail stress states

For the axial vibration and static torsion of the tubular sample for the aluminium AlMgSi alloy at the room temperature a investigations were obtained (Jakowluk & Jermolaj 1989) and Haigh's fatigue curve (Jermolaj 1992) is presented on Fig 4.1.

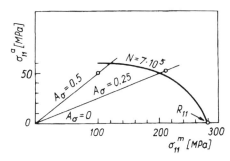

Figure 4.1. Haigh's curve fatigue for aluminium AlMgSi alloy at room temperature

The investigation results of the steel 15 HM alloy for the fatigue in biaxial stress states at the 823 K temperature (Jermolaj 1992) are presented in Fig 4.2

From Fig. 4.1. results that in the biaxial stress states vibration action on the aluminium alloy structure at room temperature, just as uniaxial stress state induces the weakening.
Instead from Fig. 2. results that in the biaxial stress states vibration action at the 823 K, just as uniaxial stress state induces the strain hardening.

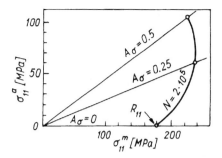

Figure 4.2. Haigh's curve fatigue for the steel 15 HM alloy in biaxail stress atates at temperature 883 K

5 MICROPOLAR WAVES AS CONVEYOR IN THE SAMPLE OF HE ATOMS AND DEFESTS

5.1 Harmonic axial loading of the sample as wave generator

The dynamical loading in the form $F_a(t) = F_a \sin \omega t$ of period τ gives the followiong positive impulse

$$I_F = F_a \int_0^{\tau/2} \sin \omega t = F_a(1 - \cos \omega t/2) = F_a/(\pi_f) \tag{5.1}$$

The positive impulse I_F enlarges mean stress σ_m.
However on the vibrocreep and fatigue efects a real influence exercises no only the positive force impulses but also the positive - negative impulse vibrations (Fig. 5.1).

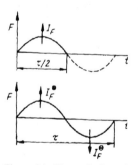

Figure 5.1. Harmonic force impulses: a) positive impulse for half a sinusoid, b) positive - negative impulse vibration.

The micropolar waves and others parameters as: σ_m, A_σ, f, T, metal alloy type have the real influence on the vibrocreep and fatigue effects.
For the show of the micropolar wave existence the investigations of the several different metal alloy creep and vibrocreep were performed (by   Jakowluk at al.). The micropolar vibrocreep strains one can to measure.

5.2 Mathematical models of the classical and micropolar waves

The motion Eq. (1.1) in the classical mechanics one can transform to the following form

$$(\lambda + 2\mu)\mathrm{grad\ div\ } \mathbf{u} - \mu \mathrm{rot\ rot\ } \mathbf{u} + \rho(\mathbf{b} - \ddot{\mathbf{u}}) = 0 \tag{5.2}$$

234

The dividing Eq. (5.2) by ρ we have the following form

$$C_1^2 \operatorname{grad} \operatorname{div} \mathbf{u} - C_2^2 \operatorname{rot} \operatorname{rot} \mathbf{u} + \mathbf{b} - \ddot{\mathbf{u}} = 0$$

(5.3)

where: $C_1^2 = (\lambda + 2\mu)/\rho, \ C_2^2 = \mu/\rho$

the quantities which two different the wave motions characterize. From Eqs. (5.3)$_2$ results that the constant C_1 the dilatancy motion characterize but constant C_2 - rotational motion. In the general case of the half - space two waves $\mathbf{u}$ and φ exist.

For the micropolar waves in the infinite micropolar elastic body the motion Eqs.(1.2) are applied, for the rotations: $\nabla \cdot \mathbf{u} = \operatorname{div} \mathbf{u}, \ \nabla \times \mathbf{u} = \operatorname{rot} \mathbf{u}$, in the following form:

$$(\lambda + 2\mu + \kappa)\nabla\nabla \cdot \mathbf{u} - (\mu + \kappa)\nabla \times \nabla \times \mathbf{u} + \kappa\nabla \times \varphi + \rho(\mathbf{b} - \ddot{\mathbf{u}}) = 0$$

$$(\alpha + \beta + \gamma)\nabla\nabla \cdot \varphi - \gamma\nabla \times \nabla \times \varphi + \kappa\nabla \times \mathbf{u} - 2\kappa\varphi + \rho(\mathbf{M} - j\ddot{\varphi})$$

(5.4)

The initial conditions in the v, for $t = 0$ are the following:

$$\mathbf{u}(\mathbf{x}, 0) = \mathbf{u}_o(\mathbf{x}), \quad \dot{\mathbf{u}}(\mathbf{x}, 0) = \mathbf{v}_o(\mathbf{x})$$

$$\varphi(\mathbf{x}, 0) = \varphi_o(\mathbf{x}), \quad \dot{\varphi}(\mathbf{x}, 0) = \nu(\mathbf{x})$$

(5.5)

The Eqs. (5.4) for the body forces: $\mathbf{b} = 0$, $\mathbf{M} = 0$ and initial conditions by Eqs. (5.5) the isotropic micropolar elastic waves (by   Parfitt and Eringen) are described the following equations:

$$\left(C_1^2 + C_2^2\right)\nabla\nabla \cdot \mathbf{u} - \left(C_2^2 + C_3^2\right)\nabla \times \nabla \times \mathbf{u} + C_3^2\nabla \times \varphi = \ddot{\mathbf{u}}$$

$$\left(C_4^2 + C_5^2\right)\nabla\nabla \cdot \varphi - C_4^2\nabla \times \nabla \times \varphi + \omega_o^2\nabla \times \mathbf{u} - 2\omega_o^2\varphi = \ddot{\varphi}$$

(5.6)

where:

$$C_1^2 = (\lambda + 2\mu)/\rho; \ C_2^2 = \mu/\rho; \ C_2^2 = \kappa/\rho$$

$$C_4^2 = \gamma/(\rho j); \ C_5^2 = (\alpha + \beta)/(\rho j); \ \omega_o^2 = C_3^2/j = \kappa/(\rho j)$$

(5.7)

For the partial reduction in Eqs. (5.6) for the regard on the vectors $\mathbf{u}$ and φ by of the scalar and vectorial potentials are presented as follows:

$$\mathbf{u} = \nabla u + \nabla \times \mathbf{U}; \qquad \nabla \cdot \mathbf{U} = 0$$

$$\varphi = \nabla\varphi + \nabla \times \Phi; \qquad \nabla \cdot \Phi = 0$$

(5.8)

Then, applying Eqs. (5.8) to Eqs (5.6) one can state that these equations are performed when:

$$\left(C_1^2 + C_3^2\right)\nabla^2 u = \ddot{u}$$

$$\left(C_4^2 = C_5^2\right)\nabla\varphi - 2\omega_o^2 = \ddot{\varphi}$$

(5.9)

$$\left(C_2^2 + C_3^2\right)\nabla^2 U + C_3^2 \nabla \times \Phi = \ddot{U}$$

$$C_4^2 \nabla^2 \Phi - 2\omega_o^2 \Phi + \omega_o^2 \nabla \times U = \ddot{\Phi}$$

(5.10)

Equations (5.9) of the scalar potentials u and φ independently are solved but vectorial potentials U and Φ of Eqs. (5.10) create the equation system. The plane wave description, of Eqs. (5.9) and (5.10), of motion of the positive versor n direction, has the following form:

$$\{u, \varphi, U, \Phi\} = \{a, b, A, B\}\exp[(ik n \cdot r - vt)]$$

(5.11)

where: a, b - complex constsnts; A, B - complex constant vectors; k = $2\pi/l$ - wave number; l - length wave; r = $x_m l_m$.
The alternately substituting of Eq. (5.11) to (5.9)₁ and (5.9)₂ the followiong plane wave displacement velocities are obtained:

$$v_1^2 = C_1^2 + C_3^2 = (\lambda + 2\mu + \kappa)/\rho$$

$$v_2^2 = C_4^2 + C_5^2 + 2\omega_o^2 k_2^{-2}$$

(5.12)

The displacement of the these waves on the base Eq. (5.11) are as follows:

$$u_1 = ik_1 a n \exp[ik_1(n \cdot r - v_1 t)]$$

$$\varphi = \nabla\varphi = ik_2 b n \exp[ik_2(n \cdot r - v_2 t)]$$

(5.13)

where: u_1 - longitudinal displacement waves and for $\kappa = 0$ these are of the classical waves. φ longitudal microrotation waves. Both type these plane waves in the vector direction n_1 (Fig. 5.2) are propagated.

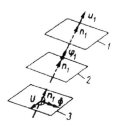

Figure 5.2. Displacement and wave microrotation: n_1 - wave propagation direction, 1 - planew wave u_1 for propagation velocity v_1, 2 - longitudinal rotation of plane wave φ_1 for v_2, 3 - transverse wave U for propagation velocity v_3, microrotaion transverse wave Φ for propagation velocity v_4.

The circular velocity of the longitudinal microrotation wave is given in Eq. (5.12). This circular velocity

$$\omega^2 = 2\pi f_2 = 2\pi v_2/l_2 = k_2 v_2$$

(5.14)

and notations (5.7)₄ - (5.7)₆, this wave velocity by Eq. (5.12)₂ one can note as follows

$$v_2^2 = \left(C_4^2 + C_5^2\right)\left(1 - 2\omega_o^2/\omega_2^2\right)^{-1} = \frac{\alpha + \beta + \gamma}{\rho j\left[1 - 2\kappa/\left(\rho j \omega_2^2\right)\right]} = (\alpha + \beta + \gamma)/[\rho j(1 - 2\omega_o^2/\omega_o^2)$$

(5.15)

Because these waves a velocity deopends from ω_2 this confirms that waves have the dissipation properties. Simultaneously for $\alpha + \beta + \gamma \geq 0$ the longitudinal microrotaion waves φ_1 can exist only for the condition

$$\omega_2 > \sqrt{2\omega_o} \tag{5.16}$$

In the cause of the critical value $\omega_2 = 2\omega_o \equiv \omega_c$ then $v_2 = \infty$ and wave no exist. THe cause $\omega_2 < \sqrt{2}\,\omega_o$ gives the imaginary velocity $v_2 = \pm i|v_2|$ an the standing wave as follows

$$\varphi = b\exp\left[-\omega_2/|v_2|\right]\mathbf{n}\cdot\mathbf{r}\exp\left(-i\omega_2 t\right) \tag{5.17}$$

The next wave solutions by (Parfitt's and Eringen's) are obtained introducing suitablely of Eq.(5.11) to Eqs.(5.10)$_{1,2}$. This givs two vectorial equations for the unknowns $\mathbf{A}$ and $\mathbf{B}$, i.e.

$$\alpha_A\mathbf{A} + i\alpha_B\mathbf{n}\times\mathbf{B} = 0, \quad i\beta_A\mathbf{n}\times\mathbf{A} + \beta_B\mathbf{B} = 0 \tag{5.18}$$

Which for the different from zero coefficients performs the resulting solutions of Eq. (5.8)$_2$ and (5.8)$_4$ in the form:

$$\mathbf{n}\cdot\mathbf{A} = 0, \quad \mathbf{n}\cdot\mathbf{B} = 0 \tag{5.19}$$

The coefficients of Eqs.(5.18) are the following

$$\alpha_A = k^2\left(v^2 - C_2^2 - C_3^2\right), \quad \alpha_B = kC_3^2$$

$$\tag{5.20}$$

$$\beta_A = k\omega_o^2, \quad \beta_B = k^2\left(v^2 - C_4^2 - 2\omega_o^2 k^{-2}\right)$$

For the conditions (5.19) results that vectors $\mathbf{A}$ and $\mathbf{B}$ lie in the one plane of the normal versor $\mathbf{n}$ to this plane. E.g. from Eq. (5.18)$_2$ we have

$$\mathbf{B} = -i(\beta_A/\beta_B)\mathbf{n}\times\mathbf{A} \tag{5.21}$$

Form Eq. (5.21) results: 1) versors $\mathbf{n}$, $\mathbf{A}$ and $\mathbf{B}$ mutually are perpendicular, 2) Both type waves are correlated because the zeroing of $\mathbf{A}$ gives also $\mathbf{B} = 0$ and from here on the base of Eqs.(5.8)$_1$ and (5.8)$_3$ one can afirm that the $\mathbf{U}$ and Φ , replying of the $\mathbf{u}$ and φ, are mutually perpendicular and to the propagation direction $\mathbf{n}$. These waves are transverse waves but these which are tied from $\mathbf{U}$ diagonal displacement waves are named, and tied from Φ - transverse microrotation waves (Fig.5.2). The first waves have the analogy in the classical mechanics in the form of the dilatation waves and in the limiting case to these waves are bringed.

6 SUMMARY AND CONCLUSIONS

The experimental vibrocreep and fatigue investigations of the steel low alloys, equally in uniaxial stress state as and biaxial stress states at the lowest temperature are shown the strongly destructive weakening vibration action on the structure, inducing the stmgth reduction and then the life.

The vibrocreep and fatigue investigations of the high-alloy steels at the high temperature, equally in the uniaxial state as and biaxial stress states the considerable strain hardening and life increase were shown.

The weakening and strain hardening phenomena are the result of the micropolar wave action along with the reflected waves cause in the sample these two so different effects.

REFERENCES

Eringen, A.C. (1968). Theory of micropolar elasticity. In "Fracture, Vol.2. Mathematical fundamentals" (Liebowitz H., ed.) Academic Press, New York and London.

Parfitt, V.R. & Eringen, A.C. (1966). Reflection of plane waves from the flat boundary of a micropolar elastic half space. Rep. No 8-3, General Technology.

Vitovec, F.H. (1957). *ASTM* **57**, p.977.

Kennedy, A.J. (1962). Process of creep and fatigue in metals. Oliver a. Boyd Ltd.

Jakowluk, A. (1967). Vibrocreep in metals (In Polish). *Publishing series "New engineering N 73"*, Warsaw.

Jakowluk, A. (1964). Certain observations on the creep of the PA3 alloy in the conditions of the static and dynamic loadings (In Polish). *2nd Symp. Pol. Soc. Mech. Th. Appl. for rheology.* Wroclaw, 51-58.

Vitovec, F.H. & Lazan, B.J. (1961). Fatigue, creep and rupture properties of heat resistant materials WADC Techn. Rep. No 561-181 in (Lazan B. *Constructive materials at high temperature; "Temperature high problems in the air design"* (In Russian), Moscow, 133-256.

Jakowluk, A. & Mieleszko, E. (1985). Construction and verification of constutive equations of the first stage of vibrocreep on the example of FeMnAl steel. *Bull. Acad. Pol.: Tech.* **33**, No 9-10, 421-428.

Sdobyrev, W.P. (1959). Creep-rupture criterion for some high-temperature alloys in a complex stress state (In Russian). Izw. *AN SSSR, Mech. & Mashinostr.* No 9, 12-19.

Jakowluk, A. (1993). Processes of creep and fatigue in metals (In Polish). Chap.5, 144-145, WNT, Warsaw.

Jakowluk, A. & Jermolaj, W. (1989). Investigation of hardening of the AlMgSi alloy in the fatigue process by means of the dynamic creep anisotropy tensor *10th Intern. Colloq. Mechan. Fatigue Metals.* Dresden, Vol.3.

Jermolaj, W. (1992). A description of dynamic creep and a failure criterion of the AlMgSi alloy in biaxial stress states. *Low Cycle Fatigue and Elasto-Plastic Behaviour of Materials* Vol.3, Elasvier Applied Science, London and New York, 375-380.

Jermolaj, W. (1992). A dynamic creep under fatigue of the 15 HM steel in biaxial stress states. *IV Inter. Symp. on Creep and Coupled Processes*, Bialystok, 109-114.

Jakowluk, A.(1990). The influence of various parameters on the process of vibrocreep and vibrocreep-rupture in different stress states in metals. *XIVth Symp. Vibrations and Waves '90*, Poznań-Blażejewko, 25-36.

Jakowluk, A., Ćech, M., Plewa, M & Jermolaj, W. (1993). Investigations of creep and micropolar creep different materials in the static and dynamic loads (In Russian). *Probl. Proćn.* Kiev, 8.

Materials development

New ferritic steels increase the thermal efficiency of steam turbines

K.H. Mayer & H. König
GEC ALSTHOM Energy, Nürnberg, Germany

ABSTRACT: Steam turbines nowadays can be operated in modern power stations at super-critical steam parameters of 600 °C and 300 bar with single and double reheat cycles. This has been made possible by extensive research into materials technology. High-temperature-resistant ferritic 9-10% CrMoV steels, which are additionally alloyed with niobium, nitrogen and tungsten, are used for the forgings and castings. Confirmation of the strength properties has been obtained in a long-term creep test of up to around max. 60,000 h for the rotor and cast steels. The strength of the turbine steels is slightly higher than that of the pipe or header steel P91 (X 10 CrMoVNbN 9 1).

The considerable experience gained by manufacturers with forged and cast large-scale components for those turbines currently being built up to 1000 MW bears witness to the favourable manufacturing properties of the newly developed ferritic steels.

1 INTRODUCTION

The steam turbines being built today for the new generation of power stations are characterized by super-critical steam parameters of about 600 °C and a live steam pressure of up to approx. 300 bar during single or double reheat cycles. Depending on the overall design of the power station, they permit improvements in thermal efficiency of approx. 6% to 8% compared with the power stations built in the 1970s and 1980s which featured steam parameters of approx. 535 °C and 185 bar (Fig. 1). Further influential factors which are used to increase the thermal efficiency

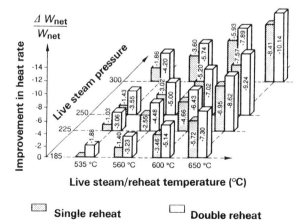

Figure 1. Process improvements by single and double reheat cycles.

of fossil-fired thermal power stations are

- improvements in the internal efficiency of the steam turbine unit,
- reduction in the exhaust steam pressure,
- increase in the feedwater pre-heating temperature in combination with a multi-stage feed heating system,
- optimization of the CRH pressure.

These measures have made a substantial contribution towards meeting the German government's target of reducing CO_2 emissions by 25%-30% by 2005 compared with 1987 levels.

The materials used for the highly stressed components of the high-pressure and intermediate-pressure turbines play a decisive part in increasing the efficiency of steam turbines, since the super-critical steam parameters can only be achieved by using materials which feature increased creep strength with sufficient resistance to oxidation and low-cycle fatigue as well as ductility and toughness. Accordingly, worldwide material research activities for fossil-fired thermal power stations focus on the further development of ferritic, high-temperature-resistant steels, since the application of ferritic steels will ensure the maintenance of the favourable operational flexibility of existing units. Fig. 2 gives a streamlined picture of the turbine development programmes in Japan, USA and in Europe. This research work has already led to the design and construction of 400 MW-1000 MW power stations of max. 593 °C / 595 °C in Japan and Europe (see Fig. 2, bottom) (Toughton 1986, Nakabayashi 1990, Scarlin 1988).

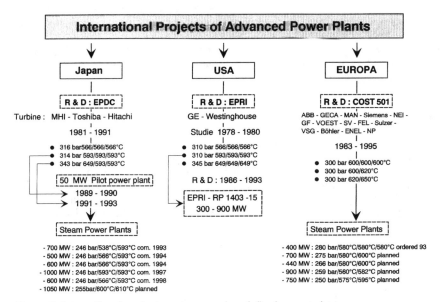

Figure 2. International projects on improved coal-fired power plant.

The work performed on material development concentrated, in particular, on increasing the creep strength of the 9 - 12% CrMoV steels by optimized alloying of tungsten, niobium, nitrogen and boron (Scarlin 1994). It is the aim of further research work to further develop the ferritic high-temperature-resistant steels in such a manner that steam admission temperatures up to 650 °C can be achieved using ferritic steels.

2 REQUIREMENTS

The competitiveness of modern fossil-fired thermal power stations compared with the existing plant and the combined gas and steam-fired power stations (GUD) means that the 600 °C turbines with greater thermal efficiencies also have to satisfy the same requirements as previous-generation steam plant, e.g.

- operating capability in the medium and peak-load range,
- service life of min. 200,000 h,
- high availability,
- long intervals between overhauls,
- short overhaul periods,
- short manufacturing times,
- competitive production costs.

These requirements demand that the use of the newly developed steels must not involve any additional risks. This means, inter alia, long creep test times to obtain reliable predictions of the creep strength for a 200,000 h period of operation, no significant tendency of the steels to succumb to long-term embrittlement as well as good manufacturing capability in hot-forming, casting and welding processes. Recent experience with the newly developed steels suggests that the manufacturing capability of the new steels is comparable with that of the older-generation steels. The ductility and toughness of the new steels are even better, despite their greater initial strength. Some uncertainties still persist in evaluating their resistance to oxidation, as the chromium content is in the 9-11% range due to optimization of the microstructure stability and thus in the lower range of the chromium content for the field-proven X 22 CrMoV 12 1 steel. Results available from oxidation tests would suggest that the lower chromium content has no significant influence on the turbine components. However, a final assessment will not be possible until practical experience has been gathered following a sustained period of operation. Counter-measures which can be exercised for functional surfaces include oxidation-resistant protective layers and hard facing. The application of high-temperature-resistant superalloys is recommended for smaller components, such as valve stems, guide bushes and highly stressed bolts.

3 CHOICE OF MATERIALS

The choice of materials for the highly stressed components is necessarily dictated by the results of international research programmes, the objective of which was to develop materials for rotors, turbine casings, valve bodies, bolts and pipes (Scarlin 1994). The rotor steels are suitable for small forgings and blades while the cast steel grades used for the casings and valve bodies are appropriate for smaller castings.

The longest creep test times over 100,000 h were hitherto achieved for the X 10 CrMoVNbN 9 1 (P91, F91) steel. The basic composition of this steel, as shown in the following table, can also be found in the new materials for rotors, casings and valve bodies.

	C	Cr	Mo	W	Ni	V	Nb	N
X 10 CrMoVNbN 9 1	.10	9	1	-	0.4	0.2	.06	.05
G-X 10 CrMoVNbN 9 1	.12	9	1	-	0.4	0.2	.06	.05
X 12 CrMoVNbN 10 1	.12	10	1.5	-	0.6	0.2	.06	.05
X 12 CrMoWVNbN 10 11	.12	10	1	1	0.7	0.2	.06	.05
G-X 12 CrMoWVNbN 10 11	.12	10	1	1	0.7	0.2	.06	.05

The C, Cr and Ni contents, which are somewhat higher in some cases, reflect the larger heat treatment diameter of the turbine rotors of 1,200 mm^2. By virtue of the additional alloying of tungsten and the higher molybdenum content of 1.5%, a further increase in creep strength against X 10 CrMoVNbN 9 1 can be expected. The longest test time attained during creep tests with the newly developed turbine steels is around 60,000 h. Fig. 3 shows the progress of tests using specimens from rotor steels compared with the creep strength of the conventional rotor steel X 21 CrMoV 12 1. Based on these figures, it is possible to make a reliable assessment of the design characteristics for 200,000 h. A comparatively good database has in the meantime also been obtained for the cast steel alternatives (Fig. 4).

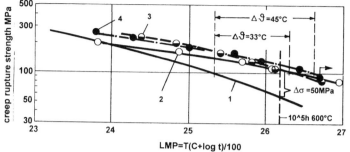

	Steel	C	Cr	Mo	W	V	Nb	N	B	Rp0.2 RT
1	X21CrMoV121	0.23	12	1.0	-	0.30	-	-	-	min. 600 MPa
2	X12CrMoVNbN101	0.12	10	1.5	-	0.20	0.06	0.05		~600 MPa
3	X12CrMoWVNbN101	0.12	10	1.0	1.0	0.20	0.06	0.05		~700 MPa
4	X18CrMoVNbB91	0.18	9.3	1.5	-	0.27	0.06		0.01	~650 MPa

Figure 3. Creep rupture strength of 10-12% Cr rotor steels.

	Steel	C	Cr	Mo	W	V	Nb	N	Rp0.2 RT
1	G-X22CrMoV121	0.00	11	1.1	-	0.25	-	-	min. 590 MPa
2	G-X12CrMoVNbN101	0.12	10	1.0	-	0.22	0.07	0.05	582 MPa
3	G-X12CrMoWVNbN1	0.12	10	1.0	1.0	0.22	0.07	0.05	592 MPa

Figure 4. Creep rupture strength of 10-12% Cr cast steels.

According to the results already gained from specimens originating from pilot components, it can be expected that the potential scatter will remain within the known range. Under an extensive joint programme of the VGB and turbine manufacturers, specimens taken from turbine rotors, turbine casings and valve bodies, which are currently being manufactured for the new generation of power plants, are being subjected to long-term investigations in order to provide statistical backing to the creep strength data of the new steels (VGB 1992). Parallel to the creep tests, extensive microstructure investigations on creep specimens tested over a long period are being performed under COST 501 so as to gain an insight for further improvement of material properties by means of optimum heat treatment and a finely matched alloying of strength-enhancing elements.

The results obtained so far from these metallographic investigations can be summarized as follows:

- the microstructures of these steels consist of sub-grains in former austenitic grains, with sub-grain boundaries, free dislocations inside the sub-grain and the $M_{23}C_6$ carbides precipitated on the phase boundaries,
- the purely thermal influence in the specimen head does not result in any change in the sub-

grain size, but do cause relatively strong growth of the $M_{23}C_6$ carbides,
- the creep deformation leads to an increase in the sub-grain size and a decrease in the dislocation density inside the sub-grain,
- the alloying of niobium and nitrogen produces fine MX precipitates (VN, Nb (CN)) with relatively high thermal stability inside the grain. These precipitates, which constitute considerable dislocation obstacles, result in the higher creep strength compared with the classic 12% CrMoV steel X 21 CrMoV 12 1,
- materials which exhibit M_2X precipitates as opposed to MX precipitates and those which possess no MX precipitates in the virgin condition feature a less stable microstructure and reduced creep strength,
- boron-alloyed versions (up to 0.010% boron) show the best microstructure stability and the highest creep strength.

In view of the medium and peak-load operating modes in which the new-generation turbines are run, the low-cycle fatigue behaviour, i.e. the behaviour in the presence of thermal fatigue stresses, is also of great interest. The results produced during the COST 501-3 programme reveal that the low-cycle fatigue strength of the new steels is similarly better than that of the steels employed in veteran power plant. Fig. 5 shows a comparison of the results for the steels

| | 28 CrMoNiV 4 9 | } | |
| - | X 21 CrMoNiV 12 1 | } | previous generation |

| | X 12 CrMoWVNbN10 11 | } | |
| - | X 12 CrMoVNbN 10 1 | } | new generation |

for tests performed at room temperature and at 550 °C. The more favourable behaviour of the new steels is shown

- in reduced temperature dependence during the tests without holding time, and

- in increased low-cycle fatigue strength with holding time influence in the long-term fatigue range, i.e over approx. 1,000 load cycles.

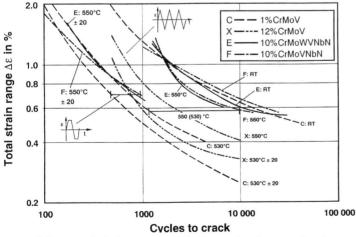

Figure 5. Low-cycle fatigue strength of advanced and conventional rotor steels at room temperature and at 550 °C (530 °C).

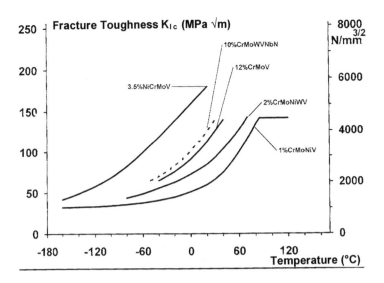

Figure 6. Fracture toughness of rotor steels versus temperature

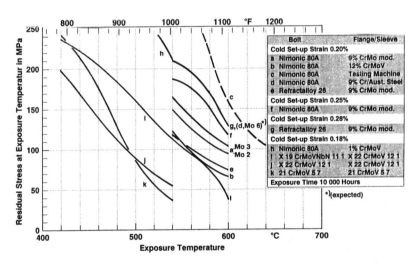

Figure 7. Relaxation strength of bolted joints for 10,000 h versus temperature

The increased fracture toughness of the new steels constitutes a further improvement of the operational properties. Both the forged and the cast components exhibit excellent fracture tougness compared with traditional steels. Fig. 6 shows this clearly with the example of fracture toughness temperature curves which were determined by means of specimens from HP, IP and LP rotors (13).

Extensive relaxation tests under the COST 501-2 programme showed that the relaxation strength of the ferritic steels at 600 °C cannot be increased substantially by optimum alloying of niobium and nitrogen or boron either. As an alternative, bolts can be used from the nickel-based alloy NiCr20TiAl (Nimonic 80A) which is also used for highly stressed bolts of the previous turbine generation in bolted joints for casings and chests made of GS-17 CrMoV 5 11. To compensate for the greater difference in thermal expansion coefficient between the 9 - 12% Cr steels and NiCr20TiAl, a greater bolt pre-loading and/or an austenitic expansion sleeve must be used - Fig. 7 (Mayer 1995).

246

The following is a summary of suitable materials for the highly stressed turbine components of modern-day power stations for steam admission temperatures of 600 °C:

Component	Material
HP and IP rotors	X 12 CrMoVNbN 10 1
	X 12 CrMoWVNbN 10 1 1
Turbine casings and valve bodies	G-X 12 CrMoVNbN 9 1
	G-X 12 CrMoWVNbN 10 1
Pipes	X 10 CrMoVNbN 9 1
Bolts	NiCr20TiAl (Nimonic 80A)
Blades, diaphragms etc.	X 10 CrMoVNbN 9 1
	X 12 CrMoVNbN 10 1
	X 12 CrMoWVNbN 10 11
	NiCr20TiAl (Nimonic 80A)

Tests on materials for steam admission temperatures of 620 °C began about two years ago under the COST 501-3 programme (Cost 501/3, 1993). The strength-enhancing effect of boron was verified by (Fujita 1978) on the "new TAF steel" which features approx. 400 ppm boron. It was found during the manufacture of pilot rotors under COST 501-1 that it is, however, only possible to control the segregation behaviour and the forgeability with a boron content reduced to about 100 ppm (Scarlin 1988). The favourable effects of about 100 ppm boron on creep strength can be seen from Fig. 3. A possible increase in the application temperature of about 20 °C to approx. 620 °C using this boron-alloyed steel can be predicted from the results available already. It is currently being investigated whether the manufacturing capability verified for rotors of 900 mm diameter can also be applied to rotors of 1200 mm diameter (VGB 1992). At the same time investigations are under way in COST 501-3 on test melts to examine the influence of increased tungsten contents and the influence of cobalt contents between 1 - 3% for boron contents of between 30 and 100 ppm. In addition to investigating the creep strength and forgeability, the aim is to gather experience on the casting and welding capability of cast steel components which, with 100 ppm boron, can only be manufactured with considerable melting and casting effort (Cost 501/3, 1993).

As yet little is known about the microstructural effect of boron on the creep strength of ferritic steels. The metallographic tests on long-term stressed creep specimens reveal thermal stabilization of the $M_{23}C_6$ carbides and a relatively stable dislocation structure. L. Lundin et al (L. Lundin) found an increase in the nucleation rate of the $M_{23}C_6$ carbides due to boron. (V. Foldyna 1995) brings up the possibility of an increase in nucleation effects for the small fine VN precipitates (MX) at dislocations.

MANUFACTURING TECHNIQUES

Based on the experience gained in the manufacture of pilot components under international material research programmes in the USA, Japan and Europe, success has now been achieved in putting the test results into operational practice by manufacturing power plant components for the new generation of turbines. The rotors are melted either according to the vacuum carbon deoxidation method or the on electro-slag remelting principle. Both methods enabled low-segregation commercial rotors featuring high levels of cleanness and fracture toughness to be manufactured from the steels X 12 CrMoVNbN 10 1 and X 12 CrMoWVNbN 10 11 with as-delivered weights of up to 38 t and max. outer diameters of 1205 mm. Figs. 8 and 9 show part of the published results of the mechanical properties obtained for pilot and commercial rotors of the

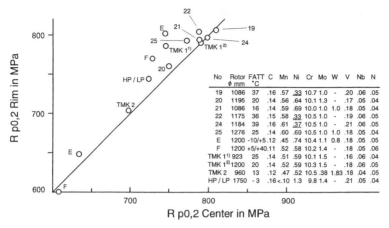

Figure 8. $Rp_{0,2}$- values as a function of rotor diameter and chemical composition of 10% CrMo(VV)VNbN steels

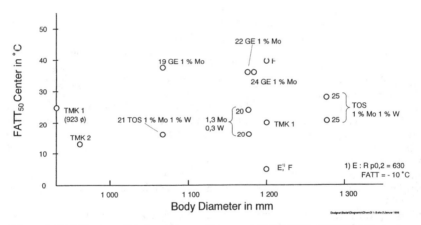

Figure 9. $FATT_{50}$ as a function of rotor diameter of 10% CrMo(VV)VNbN steels

new steels. Good through-hardening up to 1276 mm diameter has been verified for commercial rotors (Fig. 8). For a combined rotor for a steam admission temperature of max. 565 °C with a max. diameter of 1750 mm good through-hardening was also achieved by means of a higher Ni content of 1.3%; a loss of creep strength must, however, be expected with such a high Ni content. The transition temperatures for the Charpy-V-notch specimen ($FATT_{50}$) are between plus 5 and plus 40 °C, depending on the initial strength - more or less irrespective of the rotor diameter (Fig. 9).

The experience gained in Japan and Europe during the manufacture of turbine casings and valve bodies from the newly developed cast steel grades G-X 12 CrMoVNbN 9 1 and G-X 12 CrMoWVNbN 10 11 is equally positive. One major European foundry has successfully manufactured more than 50 castings (Schuster 1995).

During manufacture of the castings melting took place, as with the rotor steels, with small quantities of trace elements (As, Sb, Sn, P, S) in the electric furnace or LD converter and final alloying, desulphurization and deoxidation was handled in the ladle furnace or AOD converter. Particular significance is attached to deoxidation and the casting technique, since chrome oxides can result in critical casting flaws. The max. Al content is restricted to 0.02% so that sufficient vanadium nitrides (MX precipitates) can form. Optimum alloying of nitrogen is also important to prevent formation of pores due to non-absorbed nitrogen, As an example, Fig. 10, shows the

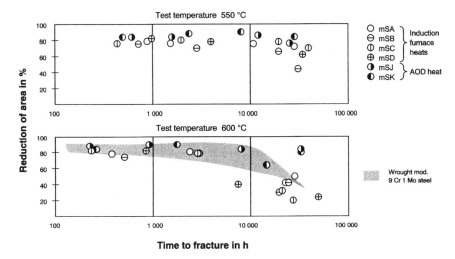

Figure 10. Creep rupture ductility of nod. 9% CrMo cast steel at 550 and 600 °C as a function of melting process

influence of the melting process on the creep ductility of the cast steel grade G-X 12 CrMoVNbN 9 1 (mod. 9% CrMo cast steel). The secondary metallurgical treatment in the AOD converter leads to a distinctly higher ductility during the creep test, both at 600 °C and at 550 °C, in comparison with melting solely in the induction furnace.

SUMMARY AND CONCLUSION

Steam turbines nowadays can be operated in modern power stations at super-critical steam parameters of 600 °C and 300 bar with single and double reheat cycles. This has been made possible by extensive research into materials technology. High-temperature-resistant ferritic 9-10% CrMoV steels, which are additionally alloyed with niobium, nitrogen and tungsten, are used for the forgings and castings. A strength analysis has been performed in a long-term creep test of up to around max. 60,000 h for the rotor and cast steels. The strength of the turbine steels is slightly higher than that of the pipe or header steel P91 (X 10 CrMoVNbN 9 1).

The considerable experience gained by manufacturers with forged and cast large-scale components for those turbines currently being built up to 1000 MW bears witness to the favourable manufacturing properties of the newly developed ferritic steels.

Current research activities are aimed in the medium term at achieving steam admission temperatures of around 620 °C and, in the long term, 650 °C - 700 °C.

REFERENCES

Foldyna V. and Kubon Z.: *"Optimized Composition of Chromium Steel for Rotor Forgings with Respect to maximum Precipitation Strengthening"* Third International Charles Parsons Turbine Conference, 25-27 April 1995, Newcastle, UK

Fujita T. et al: *"Effect of Mo and W on long-term Creep Rupture Strength of 12 % Cr Heat-Resisting Steels containing V, Nb and B"* Transactions ISIJ, Vol 18, 1978, page 115-124

Lundin L., Fällman S. and Andren H.-O.: *"Microstructure and Mechanical Properties of a 10 % Chromium Steel with Improved Creep Resistance at 600°C"*, to be published

Mayer K.H. and Neft H., 1995: *"Increasing the Efficiency of Steam Turbines by using improved 9-10 % Cr Steels"*, Conference BALTICA III, Helsinki and Stockholm, June 6-8, 1995

Nakabayashi Y. et al, 1990: *"Japanese Development in High Temperature Steam Cycles"*, COST Conference *"High Temperature Materials for Power Engineering 1990"*, Liège Belgium, Sept. 24-27, 1990

Scarlin R.B. and Schepp P., 1988: *"State of European COST Activities"*, Second EPRI International Conference on *"Improved Coal-Fired Plants"*, Nov. 2-4, 1988, Palo Alto, USA

Scarlin R.B., Mayer K.H. and Berger C., 1994: *"New Steels and Manufacturing Processes for Critical Components in Advanced Steam Power Plants"* COST 501-2 Status Report, 1994

Schuster F.A. and Cerjak H.: *"Steel Castings made from newly developed 9-12% Cr-Steels for advanced Power Generation"* 1995 ASME Cogen-Turbo Power Conference, August 23-25, 1995, Vienna, Austria

Toughton G.L., 1986: *"EPRI Improved Coal-Fired Power Plant Project"*, First International EPRI Conference on "Coal-Fired Power Plants", Nov. 9-21, 1986, Palo Alto, USA

VGD Forschungsantrag Fachausschluß - Werkstofftechnik - konventionell, Dez. 1992 *"Qualifizierung von Werkstoffen zum Einsatz in Dampfturbinen bei erhöhten Temperaturen zur Verbesserung des thermischen Wirkungsgrades und Reduzierung der Umweltbelastung"*

Europ. Forschungsprogramm COST 501/3 Runde, 1993-1997 *"Werkstoffe für hochbeanspruchte Komponenten von fossilbefeuerten Kraftwerken mit hohem thermischen Wirkungsgrad und geringer Umweltbelastung"*

Ageing of Materials and Methods for the Assessment of Lifetimes of Engineering Plant, Penny (ed.)
© *1997 Balkema, Rotterdam, ISBN 90 5410 874 6*

9 to 12% creep resistant chromium steels for high pressure piping systems in power plants

H. Weber & M. Zschau

Mannesmann Demag Energy- und Umwelttechnik, Düsseldorf, Germany

ABSTRACT: A newly developed steel X 10 CrMoVNb 91 (P 91) has been the subject of potensive research & development programms. At operating temperatures about 1010 °F (540 °C) the P 91 material offers advantages when compared to the X 20 steel commonly used in Europe. The advanced steam conditions of the new generation of power stations require this steel due to technical and economic reasons.

1 INTRODUCTION

In the early 1980's, a modified 9 % chromium steel was standardised in USA and currently identified as Grade 91 in ASTM/ASME specification (german disignation: X 10 CrMoVNb 91). The steel contains maximum 0.10 % Nb and maximum 0.25 % V in addition to 1 % Mo, which enables it to achieve sufficiently high creep rupture strength [Brühl 1989]. This steel has found to be successful material for tubing and pipework, operating at temperatures above 540 °C in fossil-fired, conventional power plants. In this temperature range this steel can bridge the gap between the ferritic steel P 22 (10 CrMo 910) and austenitic steels with respect to creep rupture strength and enables pipes with thinner walls to be used for superheaters, headers, hot reheaters and main steam pipework. Mannesmann has carried out extensive work on this steel [Brühl et al. 1989]. The intention of the following paper is first to give a general overview on the material properties and then concentrate on fabricability and welding and application.

2 PROPERTIES OF THE BASE MATERIAL

The steel X 10 CrMoVNb 91 belongs to the relatively large group of 9 to 12 % chromium steels and is mainly used in power plants for tubing (T 91), piping (P 91), as well as for forged joints (F 91), rotors and turbine casings. These steels contain molybdenum and in some cases also tungsten as main alloying elements, besides chromium. Some of them also contain niobium and vanadium, which are strong carbide forming elements.

2.1 Transformation Behaviour and Microstructure of P 91

The transformation characteristics of all high chromium steels are very similar. One of the characteristic features of these steels is that the ferrite transformation region is shifted towards longer cooling times (Figure 1). The steel develops a fully martensitic structure even if the cooling time (between 800 and 500 °C) is as long as 3000 to 4000 sec. This cooling time corresponds, for example, to the cooling rate of a pipe of 80 mm wall thickness when cooled in air. If the wall thickness is larger, faster cooling rates will be needed to avoid ferrite formation.

The transformation behaviour shown in Figure 1 is comparable to that of X 20 CrMoV 12 1. However, there are characteristic differences, too. Compared to X 20 CrMoV 12 1, the martensite start temperature is raised by about 100 K, whereas martensite hardness is reduced by about 150 Vickers numbers. Both effects are a result of the different carbon contents in both steels. The lower martensite hardness is of practical importance, because it leads to advantages in fabricability. For example the danger of intergranular stress corrosion is considerably reduced for the hardened condition after hot bending or welding. In the case of welding the danger of cold cracking is also reduced so that cooling to RT is possible.

chemical composition in %	C	Si	Mn	P	S	N	Al	Cr	Ni	Mo	V	Nb
	0.10	0.36	0.42	0.017	0.004	0.058	0.024	8.75	0.13	0.96	0.20	0.07

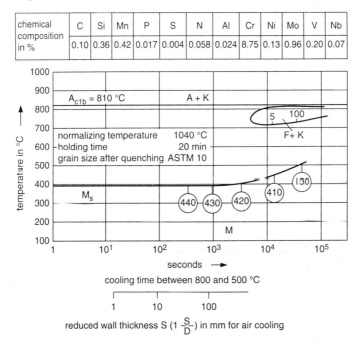

Figure 1. Cooling time temperature transformation diagram (X 10 CrMoVNb 91)

2.2 Tensile and Creep Strength

The optimum conditions of heat treatment are given by the tempering diagram in Figure 2. According to these results, the RT strength requirements of the standards are reached by 1 h tempering above 750 °C. The upper tempering temperature is limited by Ac_{1b}. Hence, the heat treatment commonly adopted consists of austenitization for 1 h at 1050 °C followed by air cooling and tempering between 750 °C and ≈ 775 °C.

Figure 3 gives a direct comparison between the creep rupture strength of X 20 CrMoV 12 1 and that of X 10 CrMoVNb 91. Compared to X 20 CrMoV 12 1, the design values for X 10 CrMoVNb 91 are higher at temperatures above 510 °C, and the difference becomes larger with increasing temperature. While the difference in design strength is 27 % at 550 °C, it increases to 53 % at 600 °C.

3 FABRICABILITY OF X 10 CrMoVNb 91

As with X 20 CrMoV 12 1, processing of P 91 by hot and cold forming as well as by welding must also be possible. It is well known in the case of X 20 CrMov 12 1 that heat treatment plays an important role in such processes. The hot bent components shall be subjected, for example, to re-normalization and tempering in order to obtain the same creep rupture strength

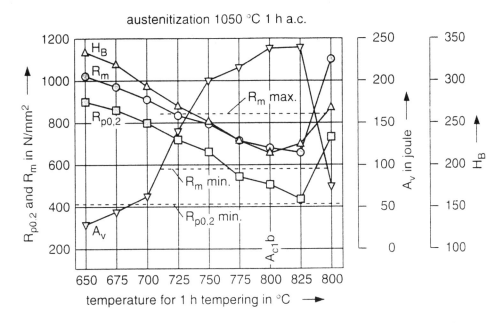

Figure 2. Tempering behaviour of X 10 CrMoVNb 91

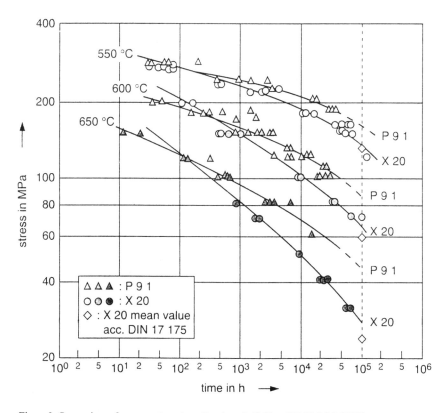

Figure 3. Comparison of creep rupture strength values for P 91 and X 20 CrMoV 121

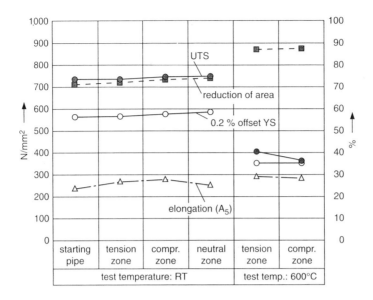

Figure 4. Tensile properties at room temperature and 600 °C of an X 10 CrMoVNb 91 induction hot bend after quenching and tempering

values as specified for piping and tubing. Owing to similar structure and transition behaviour, similar conditions for heat treatment have also to be adopted for X 10 CrMoVNb 91.

Tensile properties at room temperature and 60 °C were determined in various zones of the bend. They are compared with the properties of the initial pipe in Figure 4. The room temperature values measured in all the three zones of the bend are virtually the same and comparable to those of the initial pipe. There is also no difference in the hot tensile properties determined at 600 °C between the tension and compression zones.

4   FABRICATION OF WELDMENTS

If a new steel has to be evaluated, its response to welding has to be established. SMAW, TIG and SAW welds have been produced successfully on tubes and pipes with different wall thicknesses. The creep rupture tests on weldments made with matching and also non-matching composition consumables confirm the lower strength of the weldments (Figure 5). At stresses lower than 150 Mpa, the fracture location shifts from the base material into the softened fine grain HAZ. At 600 and 650 °C, the data points of the weldments are below those of the base material by more than 25 %. They are partly even below the average stress-rupture curve for X 20 CrMoV 12 1 base material. The results of creep rupture tests completed so far at 550 °C indicate a more favourable performance of the weldments at this temperature.

In most cases the design of component is independent of lower creep strength of welds. But at high temperature near 600 °C and above we have to analyse the stress state and to calculate the stress component normal to the weld.

Besides weldments between similar parent materials, dissimilar metal joints are of great interest. Considerable experience has been gained in the meantime, particularly in German power plants, with the long-term behaviour of welds in X 20 CrMoV 12 1 and dissimilar metal welds between X 20 CrMoV 12 1 and 10 CrMo 910.

There is, however, a basic problem, which has often been discussed in the past and led to systematic investigations [Blind, Kaes, Weber 1985]. Because of the differences in chromium contents between the materials involved, carbon diffuses from the low-chromium material into the neighbouring high-chromium steel or weld metal during the tempering treatment following

Stress, MPa

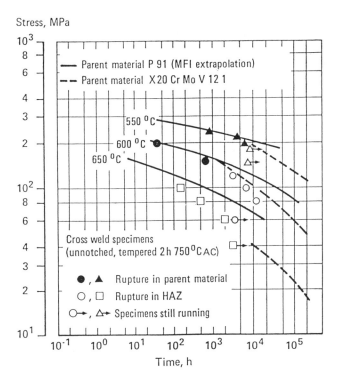

Figure 5. Creep rupture tests on weldments

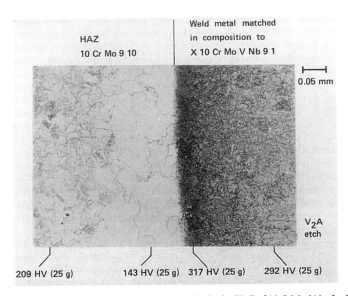

Figure 6. Decarburised grain-coarsened region in the HAZ of 10 CrMo 910 of a dissimilar joint

welding. This phenomenon results in the development of a carbon-depleted region in the low-chromium parent material or weld metal and a carbon enriched region, a so-called „carbide-band", in the high-chromium parent material or weld metal (Figure 6). The extent of these regions depends on the tempering temperature and time. It is not basically possible to avoid this

255

phenomenon, unless the welds are made with nickel-base alloys electrodes. The creep rupture tests on weldments made with dissimilar metal welds between P 91 and 10 CrMo 910 confirm the lower strength of the weldment. The data points of the weldments are below of average stress-rupture curve for 10 CrMo 910 base material (Figure 7).

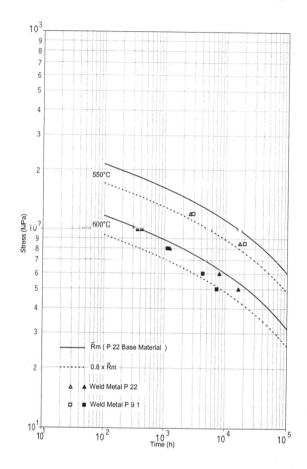

Figure 7. Creep rupture strength of dissimilar welds P 91 / P 22

The results of our weld experience can be summarised as follows. The new 9 % chromium steel P 91 is characterized by good weldability, because of its relatively low carbon content. Therefore, it should be possible to allow heavier-wall P 91 steel components to cool to room temperature after welding. The optimum post-weld heat treatment temperatures are around 750 °C, like in the case of X 20 CrMoV 12 1 steel.

No significant carbon diffusion occurs in the X 20 CrMoV 12 1 / X 10 CrMoVNb 91 joints. Consumables yielding weld metals similar in composition to either of the two base materials can be used to deposit the welds. Matching electrodes of both parent materials can also be used for dissimilar welds between 10 CrMo 910 and X 10 CrMoVNb 91. According to our latest results, rupture will occur either in the decarburised HAZ region of 10 CrMo 910 (P 91 matching electrode) or in the decarburised zone within the weld metal (10 CrMo 910 matching electrode).

The main problem associated with the weldments of this steel is the tendency of the fine grain region in the HAZ to soften, when tempered subsequent to welding as already reported in the literature [Kaes 1990]. The hardness in this region has ben repeatedly found to be about 20 HV 10 lower than that of the unaffected base material. The maximum softening was found

256

in the specimens subjected to simulated weld thermal cycles with peak temperatures in the range of 920 ± 20 °C. This softening behaviour can be confirmed by hardness measurements and rupture strength loss in slow strain rate tests. The electron microscopic examination revealed that the regions heated to these peak temperatures during welding undergo extensive recrystallization and develop large areas denuded of dislocations upon tempering at the prescribed temperatures subsequent to welding. As only a small amount of carbon goes into solution at these peak temperatures, the martensite that forms during subsequent cooling contains very little carbon and has a reduced hardness. The extent of precipitation of the coherent V-Nb carbonitride in these regions is also very much limited and the precipitates that were present coagulate so that their favourable effect on the creep rupture strength is reduced.

The softened zone in a weldment is very thin. Nevertheless, it governs the creep rupture strength of the weldment as a whole if the operating stresses are normal to the weld, because the supporting effect of higher strength neighbouring regions decreases more and more as the operating time increases and the fracture mode becomes increasingly intergranular.

5   APPLICATIONS

The question whether a certain steel can be used at specified pressure and temperature is not only dependent on material properties, but also on unit size and plant design. Therefore a comparison of applicability of different steels at different steam parameters presumes a comparable design and unit size. Thus in Figure 8 it is assumed that a certain plant can be built with 10 CrMo 910 for a main steam pressure of 185 bar and a temperature of 535 °C. If it is intended to increase the presssure to 300 bar, the steel X 20 CrMoV 12 1 has to be used. However, a temperature increase to 600 °C requires the use of X 10 CrMoVNb 91. When using X 20 CrMoV 12 1, the temperature has to be lowered by 20 K to 580 °C. According to the present technical standard it is

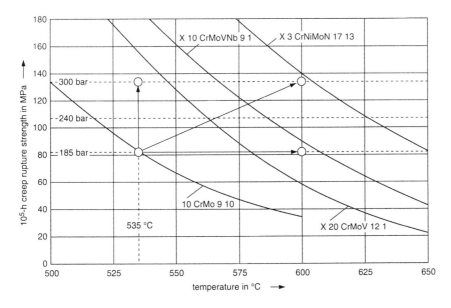

Figure 8. Requirements for creep rupture strength with increasing pressure and temperature

necessary to use the austenitic steel X 3 CrNiMoN 17 13 in order to both increase the pressure to 300 bar and the temperature to 600 °C. However, the use of austenitic steels is limited due to their unfavourable physical properties (higher coefficient of thermal expansion and lower thermal conductivity). From today's point of view, the steel X 10 CrMoVNb 91 can be used for applications involving conventional steam temperatures and pressures as well as for applications involving advanced steam parameters with higher efficiencies.

At operating temperatures around 1010 °F (540 °C), the P 91 material offers advantages when compared to the X 20 steel commonly used in Europe. The economics compared to the P 22 material are considerable.

Plant	MW	Hours of Operation	Mainsteam	Hot Reheat	Original Quality
Scholven C	400	157.000 (11.000)	540 °C/ 212 bar 1004 °F/3074 PSI	540 °C/60 bar 1004 °F/870 PSI	14 MoV 63 (1/2 Cr, 1/2 Mo, 1/4 V)
Scholven D	400	157.000 (3.000)	540 °C/212 bar 1004 °F/3074 PSI	540 °C/60 bar 1004 °F/870 PSI	14 MoV 63
Scholven E	400	145.000	540 °C/212 bar 1004 °F/3074 PSI	540 °C/60 bar 1004 °F/870 PSI	14 MoV 63
München Süd	125	130.000	540 °C/201 bar 1004 °F/2915 PSI	540 °C/59 bar 1004 °F/856 PSI	14 MoV 63
Ensdorf	300	135.000	- °C/ - bar - °F/ - PSI	530 °C/32 bar 986 °F/464 PSI	14 MoV 63
Walheim	100	150,000	540 °C/203 bar 1004 °F/2944 PSI	540 °C/47 bar 1004 °F/681 PSI	14 MoV 63
Westfalen	176	230.000	(555) °C/212 bar (1031 °F/3074 PSI) revised to 570 °C/226 bar 1058 °F/3177 PSI	- °C/ - bar - °F/ - PSI	P 22 Header 14 MoV 63 section replacement
Knepper C	300	160.000	- °C/ - bar - °F/ - PSI	545 °C/60 bar 1013 °F/870 PSI	14 MoV 63 section replacement

Table 1. Actual replacements of piping systems in Germany using material grade P/F 91

For the replacement the higher TUEV approved creep rupture strength values of X 10 CrMoVNb 91 were the basis for the design. In all these cases the whole pipework (main steam and hot reheat) were replaced. The actual operating times of the old systems were well above the original design criteria of 100 000 hrs and actually as high as 160 000 hrs. The exchange of the Chromium-Molybdenum-Vanadium-steel (DIN designation 14 MoV 63) pipe systems was necessary due to creep damage. Table 1 shows all the German power stations where this replacement took place including details like capacity, operating hours, operating conditions etc. These different projects proved that all the modified 9 % Cr-steel components required for a piping system are available in Germany and Europe (forgings, T-pieces, reducers, induction bends, elbows valves, ball pieces etc.).

All the above mentioned cases of component/system replacements based on normal (= conventional) operating parameters (T ≈ 540°C / 1000 °F; P ≈200 bar/3000 PSI). The new stations presently under construction or in the design stage show the trend to increase the operating parameters. - This is also caused by government regulations related to pollution control (especially CO_2). - Starting with 550°C (1120°F) and 286 bar (4300 PSI) (Power Station Schkopau) (Table 2) and an expected plant efficiency of 42 % the next generation of power stations will be operated at the temperature of 587°C (1090°F) and a pressure of 310 bar (4650 PSI) and an efficiency of 47 %. These are 2 stations which are presently under construction in Denmark and the high efficiency can be explained in this case to a certain extent also by the fact that seewater is being used as coolant. Table 2 summarizes the most important details of this power station in Germany and in Denmark. The capacity of these units are between 400 and 900 MW. They are coal, lignite or gas fired.

In the future the influence of the operation of power stations upon the environment will gain more and more in importance. Discussions with all parties being involved in the design, approval (authorities), location, fabrication and construction of these units prove this and the "5 Big E´s" will be the basis of all future discussions. The "5 Big E´s" are Ecology, Efficiency, Elasticity, Experience and last not least Economy.

Plant		MW	Mainsteam				Hot Reheat				Efficiency
			Temperature		Pressure		Temperature		Pressure		
Schkopau	1)	2 x 450	550 1022	°C °F	285 4132	bar PSI	565 1049	°C °F	70 1015	bar PSI	~ 40,5 %
Kirchmöser	2)3)	160	540 1004	°C °F	85 1232	bar PSI	- -	°C °F	- -	bar PSI	~ 50 %
Altbach	2)3)	330 el. 280 th.	545 1013	°C °F	285 4132	bar PSI	568 1054	°C °F	75 1088	bar PSI	~ 42 % ~ (44 %)
Schwarze Pumpe	1)	2 x 800	552 1026	°C °F	284 4118	bar PSI	570 1058	°C °F	66 957	bar PSI	~ 40,5 %
Nefo	2)	425	587 1089	°C °F	310 4495	bar PSI	580 1076	°C °F	100/30 1450/435	bar PSI	~ 47 % 4)
Skaerbaeck gas/oil or	2)	425	587 1089	°C °F	310 4495	bar PSI	580 1076	°C °F	100/30 1450/435	bar PSI	~ 47 % 4) ~ (49 %)
Boxberg	1)	2 x 800	550 1022	°C °F	285 4132	bar PSI	583 1082	°C °F	67 971	bar PSI	~ 41 %
Lippendorf	1)	2 x 800	559 1038	°C °F	285 4132	bar PSI	588 1090	°C °F	67 971	bar PSI	~ 42 %
Lübeck	2)	425	590 1094	°C °F	300 4350	bar PSI	607 1125	°C °F	70 1015	bar PSI	~ 45,7 %
Heßler	2)	700	585 1085	°C °F	300 4350	bar PSI	605 1121	°C °F	70 1015	bar PSI	~ 45 %
Frimmersdorf	1)	950	≥ 585 ≥ 1085	°C °F	≥ 286 ≥ 4146	bar PSI	≥ 605 ≥ 1121	°C °F	≥ 78 ≥ 1131	bar PSI	~ 44 %
Bexbach	2)	750	585 1085	°C °F	285 4132	bar PSI	595 1103	°C °F	70 1015	bar PSI	~ 44 %
Franken	2)3)	600 150	575 1065	°C °F	285 4132	bar PSI	595 1103	°C °F	78 1131	bar PSI	~ 43 %

1) lignite fired   2) pit coal fired   3) combined cycle   4) sea water cooling

Table 2. New powerplants/projects in Germany and Denmark using material grade P/F 91

6   SUMMARY AND PERSPECTIVES

The newly developed steel X 10 CrMoVNb 91 (P 91) has been the subject of extensive R & D programmes. Transformation behaviour and microstructure reveal great similarities with X 20 CrMoV 12 1. There are also parallels in processing by hot and cold bending as well as by welding. Besides such common features there are of course important differences resulting from the different chemical composition. While the lower carbon content is of advantage in processing and welding, creep strength is increased by the precipitation hardening resulting from the Nb addition.

An essential part of the R & D work has been to provide results to work out procedures for processing, fabrication and construction including bending and welding. The qualification of welding has been extended to dissimilar metal welds with X 20 CrMoV 12 1 and 10 CrMo 910, thereby making X 10 CrMoVNb 91 to be a candidate material not only for applications in new power plants but also for replacements during retrofitting.

Considering the creep rupture strength of heat resistant steels, X 20 CrMoV 12 1 and X 10 CrMoVNb 91 take their places within the gap formed by the low and medium alloy ferritic steels on one side and the high alloy austenitic steels on the other. The design stresses for X 10 CrMoVNb 91 are higher than those for X 20 CrMoV 12 1 in the whole temperature range that is normally used for operation. This benefit in strength can well be used in construction.

REFERENCES

Blind, D., Kaes, H., und Weber, H. 1985: Eigenschaften von betriebsbeanspruchten
Schweißverbindungen zwischen den warmfesten Stählen X 20 CrMoV 12 1 und 10 CrMo
910. *Tagungsbericht VGB-Werkstofftagung*

Brühl, F. 1989: Verhalten des 9 %igen Chromstahles X 10 CrMoVNb 91 und seiner
Schweißverbindungen im Kurz- und Langzeitversuch. *Dissertation, TU Graz*

Brühl, F., Haarmann, K., Kalwa, G., Weber, H. und Zschau, M. 1989: Verhalten des 9 %-
Chromstahles X 10 CrMoVNb 91 im Kurz- und Langzeitversuch; Teil 2:
Schweißverbindungen. *VGB Kraftwerkstechnik 69 H. 12*

Kaes, H.: Bedeutung der Wärmebehandlung beim warmfesten martensitischen Stahl X 20
CrMoV 121, *VGB Kraftwerkstechnik 70 (1990)*

Ageing of Materials and Methods for the Assessment of Lifetimes of Engineering Plant, Penny (ed.)
© 1997 Balkema, Rotterdam, ISBN 90 5410 874 6

Material development yesterday – today – tomorrow: Limitations of power plant construction

H.R.Kautz
Grosskraftwerk Mannheim AG, Germany

ABSTRACT: In order to reduce the CO_2 emissions of coal-fired power plants attempts are made during the past years to increase again the plant efficiency. This may be achieved only by raising the steam parameters, i.e. increase the operating pressure and temperature.

The desire of increasing the steam pressure and temperature for an improvement of the efficiency is not new. Around 1925, boiler and turbine manufacturers discussed the problem of being able to use the available materials only under proviso. Until this day, this has not changed very much. The problem is still how to erect a power plant with high steam temperatures (up to 700°C) and steam pressures (around 300 bar).

THE ADEQUATE MATERIAL

As previously, so today the material is the crucial item. Marguerre, founder and first chief executive of the Mannheim Central Power Plant (GKM) in 1932 raised the question, whether it was more reasonable to increase the steam pressure at steam temperatures below 500°C or to operate the system at intermediate pressures, but with a substantially higher temperature (Marguerre 1932). In those days, the answer was evident: there were (still) no materials for higher steam temperatures. Figure 1 shows the materials available in those days.

Material	Strength [kg/mm²]	Yield Strength [kg/mm²]	Chemical Analysis [weight %]			
			C	Si	Mn	Ni
Sheet plate I	35-45	19	~ 0.12	~ 0.20	~ 0.4	-
sheet plate II	41-50	22	~ 0.18	~ 0.20	~ 0.4	-
sheet plate III	44-53	25	~ 0.25	~ 0.25	~ 0.5	-
sheet plate IV	47-56	27	~ 0.20	~ 0.35	~ 0.8	-
	48-58	24	-	-	-	~ 1.5
nickel steel			-	-	-	~ 3
	65-75	45	-	-	-	~ 5

Figure 1. Materials available for power plant construction in 1927.

Figure 2 is a survey of the first high-pressure and/or high-temperature turbines which illustrates clearly the efforts made between 1925 and 1930 to raise nonetheless the steam parameters. In the USA quite early great efforts were made to achieve elevated steam parameters. In 1929, at the Trenton Channel plant of Detroit Edison experiments were conducted with steam temperatures ranging between 540 and 590°C and in 1931 - at the Delray plant - with 590°C. The material was austenitic steel.

Name of plant	commissioned	turbine manufacturer	capacity	main steam parameters	reheat	backpressure	speed	Notes
			kW	bar/°C	bar/°C	bar	s^{-1}	
Centrales electriques des Flandres, Langerbrugge, Belgium	11/1925	BBC	1650	50/4550		21	133 1/3	
Nonnendamm, Berlin	12/1925	Escher-Wyss	1000	91/400		14	166 2/3	
Gartenfeld, Berlin		Escher-Wyss	3000	181/420		37/7	100	double cylinder
Grosskraftwerk Mannheim	1928	BBC	4800	99/450		19		
Witkowitz coal mine, Karolinenschacht Czechia	1929	BBC	36000	128/500	12/360	condensation		
Langerbrugge, Belgium	1930	BBC	1000	196/450		59	116 2/3	

Figure 2. High-pressure and/or high-temperature turbines between 1925 and 1930

Austenitic steel was successfully used in the USA around 1947 in the first high-temperature power plants, and also in Germany, starting 1950, at steam temperatures around 600 to 650°C.

A publication of Waltenberger and Mattern is quoted here concerning the operational reliability of the first 600°C high-temperature steam boiler system: *"A prerequisite for the adoption of the high-temperature systems was the development of steels with an adequate creep strength and scaling resistance at high operating temperatures. The characteristics of these steels - high expansion coefficient at a reduced thermal conductivity - had to be taken into account for the design of austenitic boiler and turbine components and assemblies. In order to obtain an operational reliability a close cooperation of manufacturers and utilities was mandatory."*. (Waltenberger & Mattern 1990)

In those days, the output of power plants was low and ranged between 30 and 100 MW (Fig. 3). The plants which are planned now can hardly be compared with the old ones when considering the present capacities up to 1000 MW (Kautz & Schmidt 1996).

Figure 4 shows the present state of power plant engineering. The surprising increase of the efficiency due to elevated pressure and temperature as a result of the novel, creep resistant ferritic 9% chromium steels is clearly visible. These steels allow application at main steam temperatures up to almost 600° (Bald & Heusinger 1996).

Figure 5 is a survey of the materials used in Germany for fossil-fired power plants or intended for an application there. The steels 15 NiCuMoNb 5, 15 Mo 3, 13 CrMo 4 4, 10 CrMo 9 10, and X20 CrMoV 12 are materials used in German power plants already for many years. The steels X10 CrMoVNb 9 1 (US standards P, T, F 91) and E911, HCM 12A, and ND 616 are novel materials in power plant construction. The austenites are partly used already for a very long time (Heuser & Kautz 1996).

The necessary upgrading of lignite-fired power plants in East Germany was a decisive incentive for the development (Figure 6). Due to the transition to supercritical steam parameters in the 800 MW units which are presently under construction 40% plant efficiency are considerably exceeded (so far the East Germany plants had an efficiency of only 32 to 34%). In the figure, combined-cycle plants and circulating fluidized bed combustion may gain importance for lignite as a power plant fuel (Bald

Utility/Operator		HP steam output t/h	HP pressure bar	HP temperature °C	year
Bayer Leverkusen	boiler 1	125	160	610	1949
Bayer Leverkusen	boiler 2	125	175	650	1953
Huels, plant II	unit 1	250	340	610	1954
HEW, Neuhof plant	unit A	225	210	600	1954
Emil Adolff, Reutlingen unit 1		25	100	600	1954
Feldmuehle, Reisholz	units 2+3	65	162	605	1954
Badenwerk, Karlsruhe	unit 2	125	168	530*	1954
Harbor plant of Bremen		200	203.5	565	1954
Hannover utility	boiler	155	210	565	1954
Feldmuehle, Arnsberg plant	boilers 2+3	60	176	600	1954
Bayer, Leverkusen	boiler 3	125	175	650	1955
BASF, Ludwigshafen	boilers 10a+b	200	178	610	1955
VEW, Hattingen plant	units 3+4	316	160	605	1956
Bayer, Leverkusen	unit 4	125	175	650	1956
Badenwerk, Karlsruhe	unit 3	300	207	540*	1956
Bayer, Dormagen	unit	120	174	650	1956
Bayer, Uerdingen	2 boilers	175	330	530*	1957
Huels, plant II	unit 2	225	325	560	1958
EC, Dormagen	boilers 1+2	120	230	560	1958
BASF, Ludwigshafen	boiler 18a	210	178	610	1961
EC, Dormagen	boilers 3+4	160	230	560	1961
Bayer, Dormagen	boiler	125	174	650	1961
Bayer, Flittard plant	boiler 1	160	230	560	1962
HEW, Wedel IV plant	unit 4	570	285	545*	1962
Huels, plant II	unit 3	250	325	580	1964
Bayer, Flittard plant	boiler 2	160	230	560	1964
EC, Dormagen	boiler 5	200	230	560	1965
VEW, Schmehausen plant	unit 3	950	260	555*	1967

All data are related to superheater outlet (design)
* These boiler systems have austenitic superheater tubes only in the final superheater stage

Figure 3. High-temperature steam generators or plants in Germany (Kautz & Schmidt 1996).

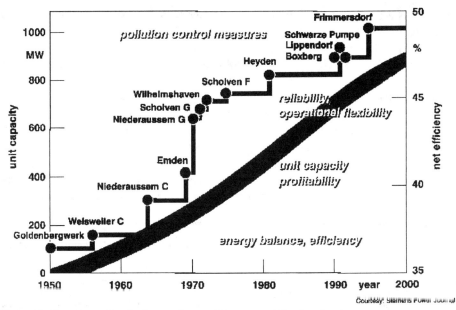

Figure 4. Increase of unit size and net efficiency of lignite-fired power plants (Bald & Heusinger 1996).

& Heusinger 1996).

The peak efficiency of the coal-fired plants presently commissioned amounts to 43%. This is based on a main steam condition of 545°C/262 bar and reheating of the steam at the transition high-pressure to intermediate-pressure turbine stage to 562°C (Single reheat). There are plans for hard coal-fired plants with partly still higher steam parameters where not only ferritic steels but also austenites are used (Fig.7) (Boehm 1994). Presently, it is impossible to state where the application limit for ferritic steels might be (Fig. 8) (Rukes et al 1994).

According to Siemens-KWU the realistic development potential for the coming decade will be a main steam condition of 630°C/310 bar, reheating also between intermediate-pressure and low-pressure at 0.04 bar. With such a design the plant efficiency would be 46%, to the last percent of efficiency increase requiring the application of austenitic instead of ferritic steel in the superheated steam region. Therefore, no final cost-benefit-assessment is possible (Bald & Heusinger 1996).

The Japanese Kawagoe power plant is worth mentioning (Kawamura et al 1991). It is a supercritical steam generator with 311 bad operating pressure and a main steam and (double) reheat temperature of 566/566/566°C. The information concerning the capacity varies slightly. In publications an output of 2150 tons/h steam is mentioned, while the present table (Fig. 9) from an international turbine development program for upgrading of coal-fired power plants reports of a 50MW pilot project. However, the design conditions for this intermediate load system are impressive. Scheduled are annually four cold starts, seven warm starts (after 48 hours), 21 hot starts (after 32 hours), 65 hot starts (after eight hours), and 1.5 rapid starts (ultra-hot starts after two hours).

ELEMENTS in weight %

DIN NAME	C	Si	Mn	P	S	Cr	Ni	Mo	Al	Cu	V	W	Ti	Nb	other
15Mo3	0,12-0,20	0,10-0,35	0,40-0,80	<0,035	<0,035	–	–	0,25-0,35	–	–	–	–	–	–	–
13CrMo44	0,10-0,18	0,10-0,35	0,40-0,70	<0,035	<0,035	0,70-1,10	–	0,45-0,65	–	–	–	–	–	–	–
10CrMo910	0,08-0,15	<0,50	0,40-0,70	<0,035	<0,035	2,00-2,50	–	0,90-1,20	–	–	–	–	–	–	–
15NiCuMoNb5	0,10-0,17	0,25-0,50	0,80-1,20	<0,030	<0,025	<0,30	1,00-1,30	0,25-0,50	<0,050	0,5-0,8	–	–	–	0,015-0,045	N <0,020
X20CrMoV121	0,17-0,23	<0,50	<1,00	<0,030	<0,030	10,0-12,5	0,30-0,80	0,80-1,20	–	–	0,25-0,35	–	–	–	–
X10CrMoVNb91	0,08-0,12	0,20-0,50	0,30-0,60	<0,020	<0,010	8,0-9,5	<0,40	0,85-1,05	<0,040	–	0,18-0,25	–	–	0,06-0,10	N 0,030-0,070
E911	0,09-0,13	<0,50	<1,00	<0,020	<0,010	8,50-9,50	0,10-0,35	0,90-1,10	<0,025	–	0,15-0,25	0,90-1,10	–	0,06-0,10	N 0,05-0,08
HCM12	<0,14	<0,5	0,30-0,70	<0,030	<0,030	11,0-13,0	–	0,80-1,20	–	–	0,20-0,30	0,8-1,2	–	<0,20	–
HCM12A	<0,15	<0,4	<0,70	<0,030	<0,020	10,0-12,0	<0,70	0,20-0,60	<0,040	<2,5	0,15-0,30	1,5-2,5	–	0,02-0,10	N 0,02-0,10
Ni616	<0,15	<0,5	<1,00	<0,020	<0,010	8,0-13,0	–	<1,0	–	–	0,10-0,30	1,5-2,5	–	<0,20	N 0,02-0,10 B <0,01
X3CrNiMoN1713	<0,04	<0,75	<2,00	<0,035	<0,015	16,0-18,0	12,0-14,0	2,0-2,8	–	–		–	–	–	N 0,10-0,18
X5NiCrAlTi3120	0,03-0,08	<0,7	<1,50	<0,015	<0,010	19,0-22,0	30,0-32,5	–	0,20-0,50	<0,5		–	0,20-0,50	<0,1	Al+Ti <0,7 Co <0,5
X8CrNiNb1616	0,04-0,10	0,30-0,60	<1,50	<0,035	<0,015	15,0-17,0	12,0-14,0	–	–	–		–	–	>10x%C	–
X8CrNiMoNb1616	0,04-0,10	0,30-0,60	<1,50	<0,035	<0,015	15,5-17,5	15,5-17,5	1,6-2,0	–	–		–	–	>10x%C	–

ferritic steels

ferritic-martensitic steels (9-12 % Cr steels)

austenitic steels

Courtesy: Deutsche Babcock

Figure 5: Chemical analysis of materials both used and envisioned

Plant	Fuel	Capacity MW	Steam Parameters[1] bar/°C/°C	Feedwater Temperature °C	Condenser Pressure bar	Net Efficiency %	Commissioned (planned)
Altbach	hard coal gas	335 (412)	249/540/565	270	0.0643	42.0 (>44)	1997
Schwarze Pumpe Units A+B	lignite	2x820	250/544/562	271	0.0485/0.0692[2]	40.6	1997/1998
Lippendorf, Units 1+2	lignite	2x931	259/550/580	271	0.039	~ 42.3	1999/2000
Boxberg, Units Q+R	lignite	2x907	258/541/578	274	0.040	41.7	2000/2002
Frimmersdorf, Unit R	lignite	1025	265/576/599	293	0.0291/0.0357[2]	> 43.0	2000
Nordjyllandvaerket	hard coal	414	285/580/580/	300	0.0235	~ 47.0	1998
Avedoere, Unit 2 (Boosting config.)	hard coal gas	375 (445)	300/580/600	310	0.022	~ 48.0 (49.5)	1999
Bexbach Unit II	hard coal	750	250/575/595	290	0.0282/0.0405[2]	46.3	2002
Hessler	hard coal	700	275/580/600	300	0.030/0.0385[2]	> 45.0	project cancelled
Luebeck	hard coal	441	268/580/600	300	0.028	~ 46.0	open
Project A	hard coal	350	273/582/595	300	0.04	> 45.0	2003
Project B	hard coal	300	300/580/600	300	0.03	~ 46.0	~ 2000
Project C	hard coal 300	300	292/577/598	300	0.03	> 45.0	open

[1] upstream of turbine   [2] condensers connected in series

Figure 6: Power plants under construction or planned [4]

266

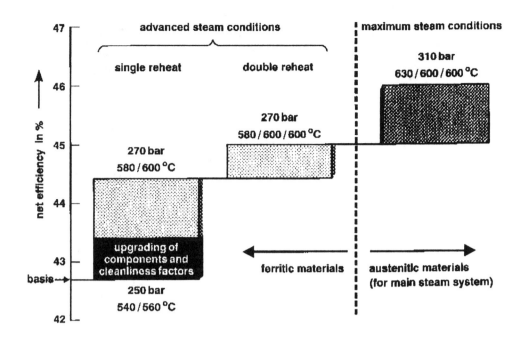

Figure 7. Effect of efficiency-increasing measured for "classic" steam power plants (Boehm 1994).

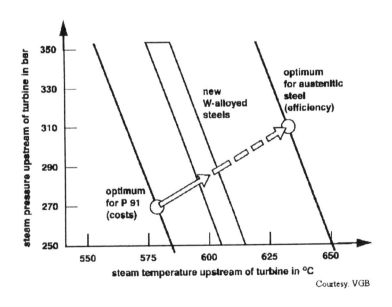

Figure 8. Limits of application for the steel P91 and austenite and estimate of evolution for tungsten-alloyed steels being developed (Reukes et al 1994).

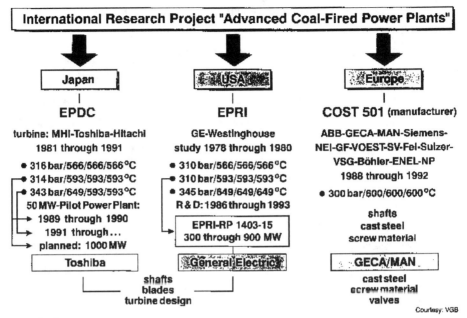

Figure 9. International research program "advanced coal-fired power plants"

The following materials were used in this power plant:

♦ for superheater tubing the austenite SA TP 347 (fine-grained);
♦ for the final superheater outlet header the already mentioned ferrite SA 335 P 91, also for the main steam line;
♦ for the casing of the high-pressure steam slide valve (Scheffknecht 1996).

Among members of the COST 501 project (European Cooperation in the Field of Scientific and Technical Research) some time ago an increase of the steam temperature in coal-fired power plants to 700°C was discussed. This would allow to reach an efficiency of approx. 55%.

In his publication "Hard coal-fired steam generators with low-NO$_x$ burners and high steam parameters Dr Scheffknecht of EVT company describes the developments thus:

"The advancement and new development of the so-called nickel-base materials open completely new perspectives. At least because of their strength values this allows to anticipate steam parameters of 375 bar/700°C. " (Scheffknecht 1996).

Figure 10 of this publication illustrates the material development and the achievable steam parameters for the steam generator wall, the waterwall tubes, and the high-pressure outlet header. The graph shows the time from which the various materials are likely to be available and also the steam parameter limits classified by the above mentioned components (Scheffknecht 1996).

For decades, the materials and the weldability decided the limits of the steam parameters and the power plant capacity. Here, the chemical high-pressure industry was frequently pioneer, also with respect to pressure and temperature. Foundries,

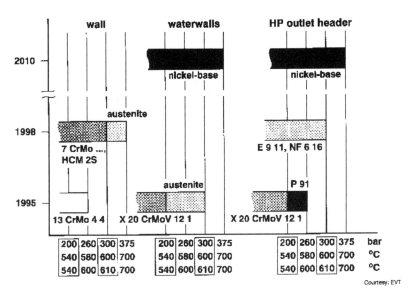

Figure 10. Materials development time and achievable steam parameters for the waterwall, superheater tubes, and HP outlet header.

steel mills and forging shops with expertise in manufacturing power plant components were already available at an early stage thanks to the requirements of the chemical industry. When looking back, the problems with drums (weight on boiler structure or manufacturability - riveted, forged or welded drums) resulted after 1925 in the boiler without drum, the Benson design. The difficulties with the necessary (ultrapure) boiler water ('Benson sickness') evolved into what is known today as state of the art cycle chemistry. As a result of the development in the field of welding methods and fillers the difficulties in welding the monotube steam generators could be overcome. Novel design features and the welding techniques allowed to increase the plant capacity, last but not least as a result of the application of the "fin tube walls". Power plant engineering took many detours (which appear very peculiar today) in order to obtain excellent technical solutions. The present state of power plant engineering demonstrates the success.

In order to obtain this level a close cooperation between manufacturers and utilities was required, as already mentioned.

Actually, the first high-temperature power plant (Waltenberger & Mattern 1990), and also the later systems in the chemical industry, were thus a 'coproduction' of the Farbenfabriken Bayer company as the planning party, the Duerr company at Ratingen, the Siemens-Schuckert company at Muelheim/Rohr (as turbine manufacturer and the Deutsche Mannesmann-Roehrenwerke at Dusseldorf (pipe supplier).

What is lacking today for the manufacture of such boiler system is the indispensable infrastructure. Who shall melt the required raw material? Who can machine it? How is austenite formed in order to obtain the adequate grain size in the component material so that ultrasonic examinations are possible? The temperature window for the planned nickel-base materials is very narrow for forming and subsequent heat treatment.

269

Phase	Duration	Subject
1	3 years (1996-1998	Feasibility study: * Clarification of technical and economic possibilities (design, thermodynamics, investment costs * Optimization of overall plant process * Identification of critical components, dimensions, loads, material requirements * Review of application/properties of available materials
2A	6 years (1999-2004)	Material development, optimization and testing
2B	4 years (1999-2002)	Design and manufacture of test items of "critical" boiler and turbine components
3	5 years (2002-2007)	Trial operation of "critical" components in conventional power plants - compilation of experience
4	3 years (2006-2010)	Design and erection of demonstration plant
5	5 years (2011-2013)	Erection and trial operation of demonstration plant

Figure 11. Development project THERMIE 700°C.

Who will undertake the component manufacture? Fig. 11 lists the requirements.
And which utility will venture to order and operate such a power plant? Where are the experts from the various fields of material science, of machining, welding and testing? Where are the experts in the utilities whose support is mandatory for the erection and operation of such a plant? Who will establish the infrastructure required for the construction of a few of those plants?

When looking at the previous evolution of power plants it is clear that a maximum-temperature power plant could be erected. Maybe, the steam parameters and the capacity will be slightly lower than planned now.

CONCLUSION

The improvement of the thermodynamic cycle made the application of austenitic steels uneconomic at the beginning of the sixties because of the high material costs. The martensitic steel X20 CrMoV 12 1 (Fig. 5) known since 1925 had matured enough by 1955 to be applied. It proved to be successful and reliable until this day.

The transition to ferritic steel caused the loss of knowledge and expertise in the manufacture of austenitic forgings. This will be sorely felt when erecting the future high-temperature power plants where again austenitic or nickel-base materials are intended for use.

The increase public awareness of environmental pollution resulted in discussions of raising the steam parameters in power plants again. Novel ferritic-matensitic high-temperature steels (e.g. P 91, NF 616; Fig 5) were developed which allow to reach steam temperature up to or around 600°C or even more. The development project THERMIE 700°C - a European project - reaches still further. Until the year 2010/2015 the temperature in coal-fired power plants shall be increased to approx. 720°C and the plant efficiency to 55% (Figure 11). This project is based on nickel-base alloys and superalloys. The imponderabilities are listed in Fig. 12.

Imponderabilities, risks, and possible impediments until the realization of optimized power plants:

- Large-scale availability of adequate apparatus and turbine materials for high temperatures (ferritic, austenitic) must exist at justifiable costs.
- Operational compatibility of maximum furnace exit temperatures of 1000 to 1050 °C must be verified. Measures for e.g. monitoring of coal quality and combustion characteristics in the furnace may be required to preclude caking and slagging.
- The unit size is limited for technical and economic reasons.
- Availability and costs of adequate waste gas heat utilization systems must be verified ($T_{waste\ gas} < 110\ °C$).
- The smooth interaction of new components with respect to plant safety, availability, and operation must be demonstrated.
- Any further improvement of the cold end is limited by the system condenser-cooling tower.

Figure 12. Imponderabilities, risks, possible impediments for realization of the optimized power plant.

REFERENCES

Bald, A. and Heusinger, K. "Energy generation with advanced steaming plants will relieve the environment" (in German). *Siemens Power Journal* (1966), pp 5-11.

Boehm, H. "Fossil-fired power plants" (in German). *VGB Kraftwerkstechnik 74.* (1994), pp 173-186.

Heuser, H. and Kautz, H.R. "Martensitic and austenitic materials for high-temperature power plants - design, manufacture and operational problems" (in German). Schweissen und Schneiden (1996), DVS Berichte Bank 175, pp 158-166.

Kautz, H.R. and Schmidt, P. "Are power plants with steam parameters of 700°C and 300 bar feasible? VDI Berichte 1300, *Conf. Material - Component Damage* (1966), pp153-181.

Kawamura, T. et al. "Planning and operation of supercritical steam generators with 311 bar in the Kawagoe power plant". *VGB Kraftwerkstechnik 71* (1991), pp 637-643.

Marguerre, F. *High steam parameters* (in German). Z.VDI, March 1932, pp 287-292.

Rukes, B., Vollmer, W., and Wittchow, E. "Power plants with high steam parameters" (in German) *VGB Kraftwerkstechnik 74* (1994). pp 405-411.

Scheffknecht, G. "Coal-fired steam generators with low-NO$_x$ burners and high steam parameters" (in German). *75 years KSG-EVT* (1996), pp 25-33.

Waltenberger, G. and Mattern, P. "Application of austenitic materials in steam boilers - operational approval in the first 600°C high-temperature boiler system" (in German). *VGB Kraftwerkstechnik 70* (1990), pp 68-76.

Ageing of Materials and Methods for the Assessment of Lifetimes of Engineering Plant, Penny (ed.)
© *1997 Balkema, Rotterdam, ISBN 90 5410 874 6*

The susceptibility of low temperature sensitization of 304 S.S. after mechanical stress improvement process (MSIP)

K.Y.Hsu
Materials Research Laboratories, ITRI, Taiwan

ABSTRACT: Mechanical Stress Improvement Process (MSIP) shifts the stress state of the interior surface of a pipe weldment from tensile to compressive which mitigates IGSCC of stainless steel piping in BWR's. After the MSIP permanent plastic strains remain under the MSIP tool region. During service at 288°C BWR environment, it is possible that the plastic strain may induce low temperature sensitization (LTS). In this work, AISI 304 S.S. was cold rolled to various degrees of reduction of thickness to simulate plastic deformation produced by MSIP. Cold rolled specimens were then subjected to LTS treatment at 500 °C for 24 hrs. The influence of plastic deformation on LTS was studied using the ASTM standard A262 practices A and E, the electrochemical potentiokinetic reactivation (EPR) technique and constant extension rate test (CERT). With increasing degree of cold rolling, the Cr depletion zones mostly concentrated on slip bands, as observed in the results of ASTM A262 practices A and E , and EPR test. CERT data indicated in simulating BWR water IGSCC would not occur in plastically deformed 304 S.S. when subjected to LTS heat treatment.

1 INTRODUCTION

Although the effectiveness of Mechanical Stress Improvement Process (MSIP) to produce a desired stress state has been verified by residual stress measurement and finite element analysis, there are few studies on the effect of permanent plastic deformation remaining at the MSIP tool region. Low temperature sensitization (LTS) refers to sensitization at temperatures below the typical range of sensitization (500°C - 800 °C). Earlier studies (Povich 1978a & 1978b) indicated that (1) a prerequisite for LTS phenomena is the presence of chromium carbide nuclei at the grain boundaries and (2) during LTS no new nuclei form, but existing nuclei grow. Majidi and Streicher (Majidi 1984) proposed that (1) preseeding of grain boundaries carbide nuclei is not a prerequisite for LTS to occur and (2) prior cold work introduced in the surface layer by cutting or grinding can cause favorable thermodynamic and kinetic conditions for carbide nuclei to form and grow during LTS. They also reported the prior 5 and 20 % cold work results in LTS (500 °C for 24hr) with precipitation of chromium carbides almost entirely confined to slip lines. Shah (Shah et al. 1990) reported for a given composition there was a threshold cold work level for inducing LTS (500°C for 24hr) at grain boundaries but a higher level of cold work (about 30%) can prevent LTS by nucleating carbides within the grains. From these studies, there is a discrepancy concerning the precipitation site of carbides. In the present work, the AISI 304S.S. was cold rolled to various degrees of reduction of thickness and subsequently subjected to LTS heat treatment at 500°C for 24hr. An attempt was made to resolve the discrepancy by using the ASTM standard A262 practice A and E, the electrochemical potenoikinetic reactivation (EPR) techniques. The stress corrosion cracking (SCC) susceptibility of the LTS pecimens was investigated by means of constant extension rate tensile (CERT) tests in simulated boiling water reactor (BWR) environment.

2 EXPERIMENTAL PROCEDURES

AISI 304 S.S., whose nominal composition is shown in Table 1, was used in this investigation. The as received material in mill annealed condition was subjected to solution heat treatment at 1050 °C for 1hr followed by water quench. Sheets of 200 x 70 x 10mm were cold rolled at ambient temperature to give various amounts of reduction in thickness, namely 5, 10, 30, and 50%. The cold rolled coupons were subjected to LTS at 500°C for 24hr in air furnace. Based on reported activation energy of ~ 146 - 167 kJ/mol $^{-1}$, this LTS corresponds to about 10 years of service at 300°C (Shah 1990). In each cold rolled and LTS treated coupon, two samples were cut, one parallel to rolling direction (L) and the other transverse (T), for performing ASTM standard A262 Practice A test and the EPR test. The balance of the coupons was used for ASTM A262 Practice E tests and CERT tests after polishing successively on silicon carbide emery paper up to a final grade of 600 grit.

Table 1. Chemical Composition of AISI 304 S.S. (wt%)

C	Si	Mn	P	Cr	Ni	Mo	Fe
0.05	0.5	1.52	0.02	18.5	9.6	0.14	bal.

2.1 Oxalic Acid Etch Test (ASTM A 262 A)

All the cold rolled and heat treated specimens were electrolytically etched in 10% oxalic acid solution for 1.5min at a current density of 1.0 A/cm^2. The specimens were given a prior 1 μm polishing treatment . Evaluation was conducted by microscopic examination and classification of the etch stucture as step, dual, and ditch for IGA screening.

2.2 Copper - Copper Sulfate -16% Sulfuric Acid Test (ASTM A262 E)

The ground specimens were exposed to boiling 16% sulfuric acid containing metallic copper for 72hr. After tests the specimens were bent into a U-shape and examined on the outside surface at 40x for fissures.

2.3 Double Loop EPR tests

For the double loop EPR tests, the procedure described by Akashi, et al (Akashi 1980) was used. The test solution was 0.5 M H$_2$SO$_4$ with 0.01M KSCN and deaerated by pure nitrogen gas at temperature of 30 °C. The corrosion cell consisted of a 1L five-neck flask with high density graphite as counter electrode and saturated calomel electrode (SCE) as reference electrode. The polished and mounted specimen was immersed in the test flask for about 2 min, then polarized anodically to a potential of +300 mV VS SCE at a rate of 6V/hr. As soon as this potential was reached, the scanning direction was reversed to the free corrosion potential. The maximum current for each loop was measured : Ia for the large anodic loop and Ir for the smaller loop generated during reactivation. The ratio of Ir/Ia was used as a measure for the degree of sensitization.

2.4 CERT Test

The 3mm-thick plate specimen with a 18-mm gauge length was used in CERT test. Each specimen was tested individnally in an autoclave with a static water system, which contained high purity water (<0.5μS/cm). In each run, the autoclave was firstly aerated at room temperature for 1hr to have water oxygen saturated, then the temperature was raised to 288°C with an pressure equilibrium vapor pressure of 7.48MPa. The tests were conducted at a strain rate of $5 \times 10^{-7} s^{-1}$ After the tests were completed, the ultimate tensile stress and total elongation were recorded, In addition , the fracture surfaces were examined by scanning electron microscopy (SEM).

3 RESULTS

3.1 ASTM A262 A & E and DL EPR test

The results obtained in EPR and ASTM A262 plactice A and E tests for 5 to 50% cold rolled 304S.S. specimens are shown in Table 2 & Table 3. Table 2 gives tests results of cold rolled specimens, and Table 3 are test results of cold rolled with LTS treatment specimens. Microstructures observed after oxalic acid etching are shown in Figure 1. The results of bending test after ASTM A262E test are given in Table 2 & 3 . Microstructures produced after DL EPR test are shown in Figure 2. From the DL EPR experiments, Ir/Ia was calculated which is the ratio of the peak current during reactivation to that during activation as shown in Figure 3. The values are also given in Tables 2 & 3.

3.2 CERT Test

Results of the CERT tests carried out in simulated BWR water are summarized in Figure 4, which gives UTS and total elongation vs. cold rolling. Representative SEM micrographs are shown in Figure 5.

Table 2. Results obtained in EPR experiments and ASTM A262 practice A and E (cold rolled specimens)

Specimen / Test Method	As Received	Cold Rolling(%)							
		5%		10%		30%		50%	
		L^1	T^2	L	T	L	T	L	T
DL EPR(Ir/Ia)	10^{-3}	4.2×10^{-4}	3.9×10^{-5}	10^{-4}	2.8×10^{-5}	1.1×10^{-3}	8.2×10^{-4}	8.3×10^{-5}	9.1×10^{-5}
ASTM A262 A	Step	Step	Step	Step	Step	Step	Step	Step	Step
ASTM A262 E	NA^3	NA	NA	NA	NA	NA	NA	NA	NA

L^1:Rolling Direction
T^2:Trans. Direction
NA^3:No Attack

Table 3. Results obtained in EPR experiments and ASTM A262 practice A and E (cold rolled +LTS specimens)

Specimen / Test Method	As Received	Cold Rolling(%)+500°C/24hr							
		5%		10%		30%		50%	
		L	T	L	T	L	T	L	T
DL EPR(Ir/Ia)	4×10^{-3}	0.091	0.085	0.255	0.298	0.403	0.379	0.453	0.48
ASTM A262 A	Step	EIG^1	EIG	EIG	EIG	EIG	EIG	EIG	EIG
ASTM A262 E	NA	LA^2	LA	A^3	A	B^4	B	B	B

EIG^1 :Etch inside grain
LA^2 :Lightly Attack
A^3 :Attack
B^4:Broken into two pieces

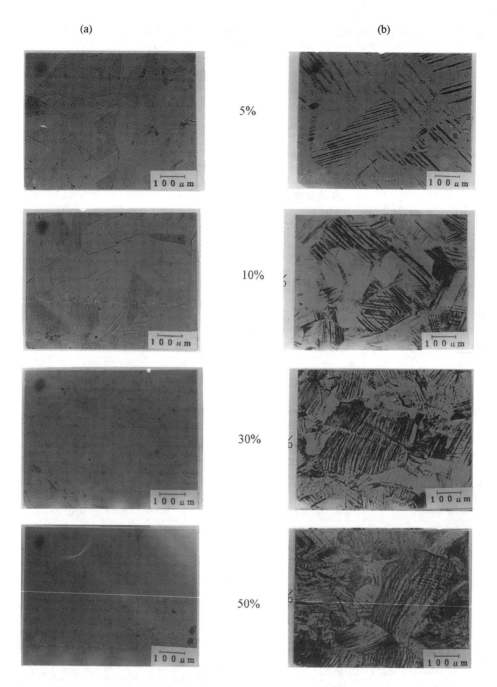

(a)   (b)

5%

10%

30%

50%

Figure 1. Microstructure of 304 S.S. after oxalic acid etch test (a) with 5%,10%,30% and 50% cold rolling. (b) with 5%,10%,30% and 50% cold rolling + LTS.

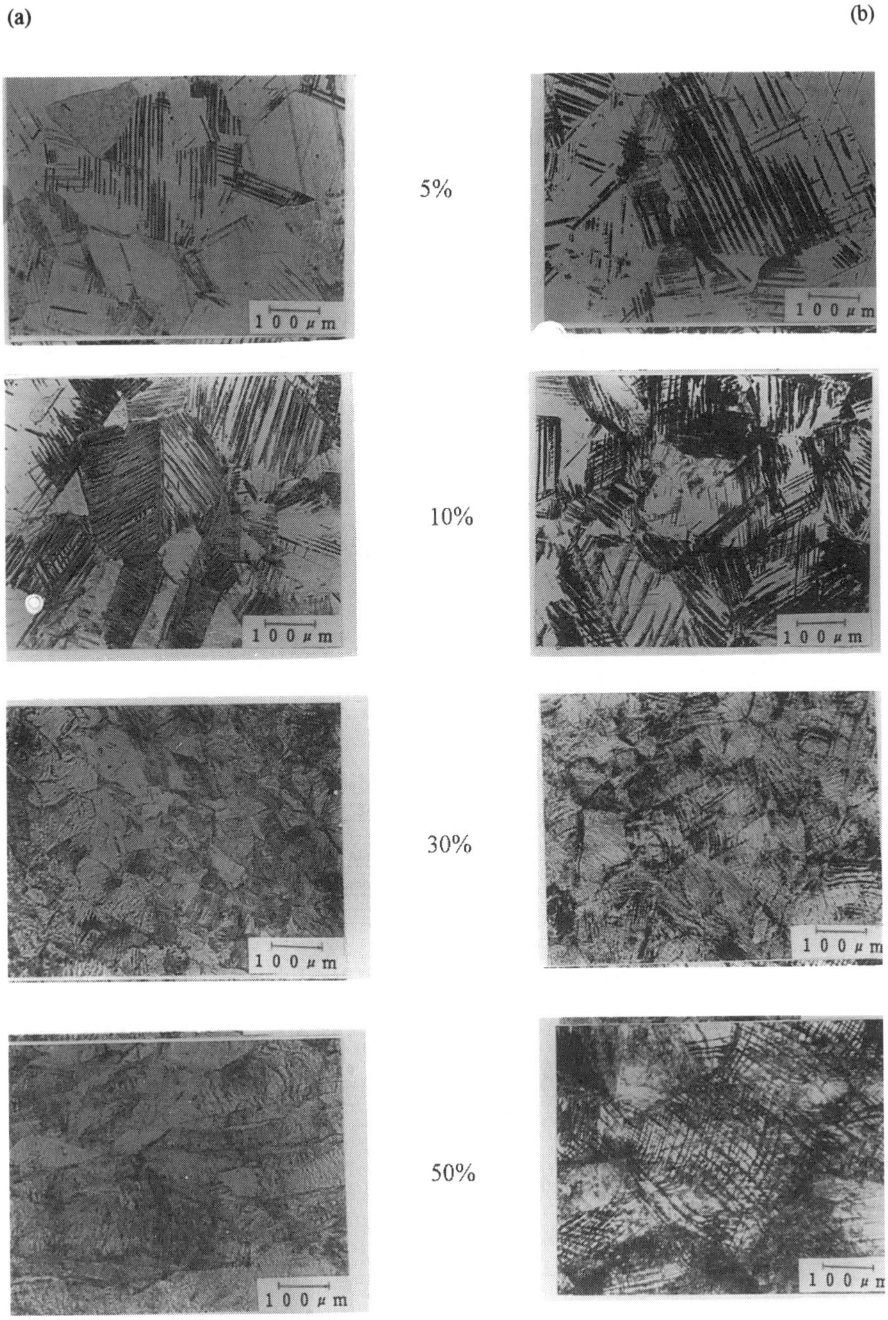

(a)　　　　　　　　　　　　　　　　　　(b)

5%

10%

30%

50%

Figure 2. Microstructure of 304 S.S. after DL EPR test (a) with 5%,10%,30% and 50% cold rolling. (b)with 5%,10%,30% and 50% cold rolling + LTS.

277

DL EPR TEST(5T1BH1)

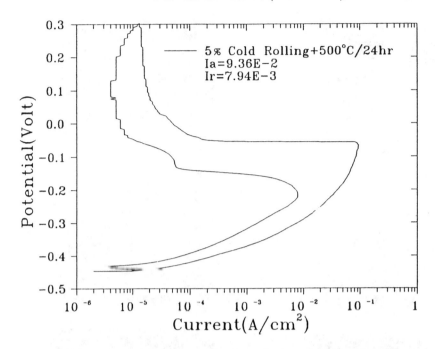

Figure 3. Anodic polarization curve showing activation and reactivation in DL EPR for 5% cold rolling +LTS specimen.

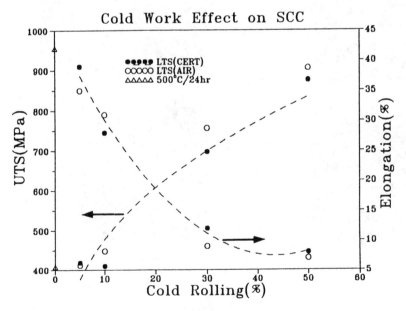

Figure 4. SCC behavior of 304 S.S. variously cold rolled + LTS in simulated BWR water.

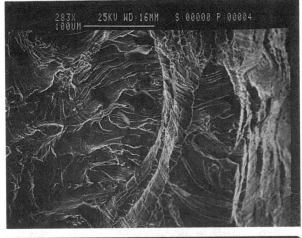

(a)

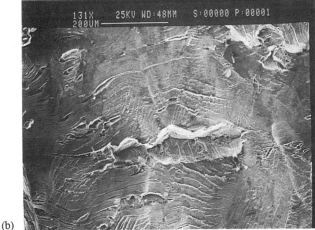

(b)

Figure 5. SEM fractographs of 5% cold rolling + LTS after SCC test
in simulated BWR water. (a)fracture surface (b)gauge length

4 DISCUSSION

4.1 *ASTM A262 A and E Tests*

The cold rolled specimens always show a "step" structure (Figure 1a), and the slip bands increasing with higher level of cold rolling are slightly etched. The cold rolling + LTS specimens have a "step" structure at grain boundaries, but the slip bands produced by cold rolling are heavily etched, as shown in Figure 1b. It seems that the precipitation of chromium carbides is confined almost entirely to slip bands with substantially no precipitation at the grain boundaries.

It is known that AISI 304S.S. undergoes martensitic transformation with the high degree of cold working (Elayaperumal 1972). In this work, ferrite measurements using a ferrite meter indicated that the 10% cold rolled specimen contained only a negligible amount of ferrite, whereas the 30% cold rolled specimen contained ~5% ferrite. Based on the results of the oxalic acid tests it can be remarked that the slip bands and martensite, both induced by cold rolling, accelerate the precipitation of chromium carbides inside the grains, as they increase the number of nucleation sites and facilitate the diffusion processes in the matrix.

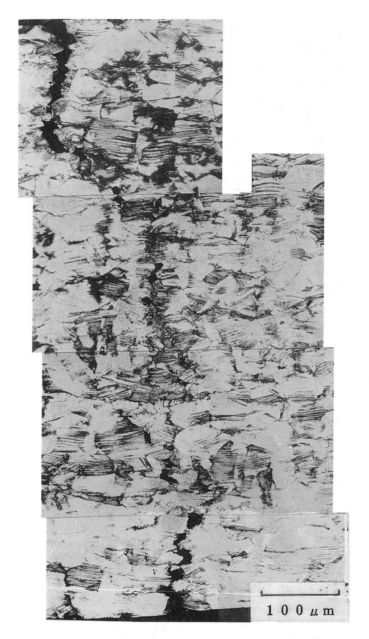

Figure 6. A penetrated crack found in 30% + LTS specimen after Strauss test.

The results of Strauss test shown in Table 2 & Table 3 indicate all cold rolled specimens have no crack following bending, while cold rolling + LTS specimens exhibit different behavior according to various extent of prior cold rolling. The 5% cold rolling + LTS specimen has surface fissures after bending and the 10% cold rolling + LTS specimen has deep cracks (approximate 1/3 of thickness). The 30% and 50% cold rolling + LTS specimens break into two pieces after bending. The surface examination after Strauss test shows , as in oxalic acid tests, slip bands inside grains are severely attacked in LTS specimens with increasing extent of cold rolling. A crack penetrating through thickness of specimen was found in 30% + LTS specimen before bend-

280

ing, as shown in Figure 6. Owing to the severe environment during Strauss test, the slip bands attacked inside individual grain were linked together which resulted in penetrating crack.

4.2 *DL EPR Test*

The ratio of Ir/Ia for various cold rolling and cold rolling + LTS specimens after DL EPR test is shown in Figure 7. The classifications obtained in the oxalic acid etch test (ASTM 1978) for non-cold working AISI 304S.S are also shown in Figure 7. The ratio of Ir/Ia for cold rolling specimen are always below 0.001 (as shown on Table 2 & Table 3), that is , correspondent to "step" structure in oxalic acid etch test. On the other hand, the Ir/Ia values of LTS specimens increase with increasing higher level of prior cold rolling and all of the Ir/Ia values lie on "Ditch" structure. Microstructure inspection after DL EPR (Figure 2) indicates cold rolling results primarily in chromium carbide precipitated at slip bands rather than at grain boundaries during LTS treatment. As the cold work increases, the AISI 304 S.S. will induce matensite phase. It is interesting to know if any microstructure change of AISI 304 S.S. will affect the sensitivity of DL EPR test. Figure 8 shows peak current of anodic loop for cold rolling and cold rolling + LTS specimens during DL EPR scan. With increasing extent of cold rolling, whether subjected LTS or not, the peak current of anodic loop is approximately equal to a constant value (about 9×10^{-2} A/cm^2). During anodic scan, any different phase (such as martensite) or microstructure defects (dislocation and slip bands) of specimens are all in active state with very high dissolution rate, which contributes almost equally to overall peak current Ia. Detecting the susceptibility of sensitization depends on reactivation scan which depassivates the chromium depleted regions causing their dissolution while the passive film elsewhere remains intact.

From the above observation, it is clear that the degree of a sensitization of cold rolled + LTS AISI 304S.S. can not be determined by the DL EPR test because the attack entirely concentrates at slip bands. In addition, the results of DL EPR is consistent with that of ASTM A262 A&E.

4.3 *CERT Test*

Figure 4 shows that with increasing cold rolling the UTS increases and total elongation decreases after CERT test. There is a little reduction of UTS compared with air test data. None of the cold rolling + LTS specimens shows an intergranular crack. Fig 5 shows an SEM fractograph of 5% cold rolling +LTS specimen. As can be seen (Figure 5a), the fracture is transgranular only for a short distance from the edge and is almost ductile over the rest of fracture surface. In addition, there are many secondary cracks distributed across the gauge length, which are initiated at the slip bands (Figure 5b). The same behaviors can be found on the specimens of 10%, 30% and 50% cold rolling + LTS. It is also found that the initial transgranular depth decreases as cold rolling increases.

The fact that CERT data were inconsistent with those of ASTM A262E can be attributed to the following : (1) Relative to ASTM A262E test, the environment of CERT test is very mild. The pure water in autoclave is stagnant, there is no fresh water added in. (2) When crack initiates, the local strain rate at the crack tip may be large though the nominal strain rate is 5×10^{-7} s^{-1}. (3) According to slip dissolution / film rupture model, the slip steps which are dissolved and then passivated must be continuously generated to induce crack advance. In the cold rolled structure, there are dense and crossing slip bands in each grains, which may impede dislocations moving to form slip steps when subjected to straining.

5 CONCLUSIONS

From the above mentioned, the following conclusions can be drawn:
(1) For AISI 304S.S., cold rolling as low as 5% can accelerate the chromium carbide precipitation at slip bands after 500°C/24hr heat treatment. The effect is more evident when martensite is present.

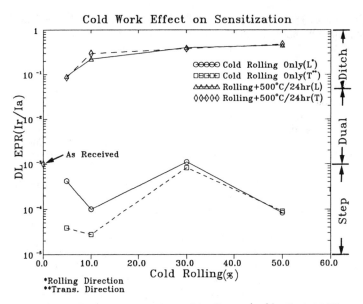

Figure 7. The ratio of Ir/Ia for various cold rolling and cold rolling + LTS specimens after DL EPR tests.

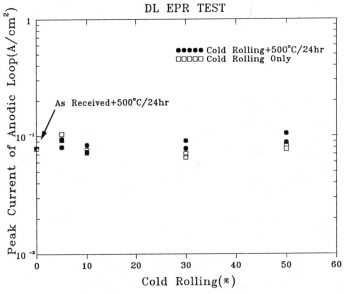

Figure 8. Peak current of anodic loop for cold rolling and cold rolling + LTS specimens (DL EPR test).

(2) Good correlation between DL EPR test and the ASTM A262E test could be obtained for cold rolling +LTS 304S.S.

(3) For cold rolling + LTS 304S.S., current ratio Ir/Ia of DL EPR test could not be used as an index of grain boundary sensititation.

(4) CERT data in simulated BWR water indicate IGSCC would not occur in plastically deformed 304S.S. when subjected to LTS heat treatment.

REFERENCES

Povich, M. J. 1978a. Low Temperature Sensitization of Type 304 S.S. *Corrosion*, 34(2): 60-65.

Povich, M. J. & Rao, P. 1978b. Low Temperature Sensitization of Welded Type 304 Stainless Steel. *Corrosion*, 34(8):269-275.

Majidi, A. P. & Streicher, M. A. 1984. The Effect of Methods of Cutting and Grinding on Sensitization in Surface Layers on AISI 304 S.S.*Corrosion*, 40(9): 445-458.

Shah, B. K. et al. 1990. Effect of prior cold work on low temperature sensitization susceptibility of austenitic stainless steel AISI 304. *Material Science and Technology*, 6: 157-160.

Akashi, M. et al. 1980. *Corrosion Engineering*, 29: 163.

Elayaperumal, K. et al. 1972. Passivity of type 304 stainless steel-effect of plastic deformation. *Corrosion*, 28(7): 269-273.

ASTM, 1978. *ASTM Book of Standards* , Vol. 10, A262. ASTM, Philadelphia, Pennsylvania.

Ageing of Materials and Methods for the Assessment of Lifetimes of Engineering Plant, Penny (ed.)
© *1997 Balkema, Rotterdam, ISBN 90 5410 874 6*

Fitness for service considerations: The Meyer hardness test applied to cold rolled and annealed steel to analyse its physical state

Ph.Tipping
Swiss Federal Nuclear Safety Inspectorate (HSK), Villigen, Switzerland

V.Levit
Federal University of Pelotas, RS, Brazil

ABSTRACT: The influence of time at temperature can, potentially, affect the mechanical and corrosion properties of metal alloys due to phase changes, precipitation of intermetallic compounds, diffusion of impurities or alloying elements and grain growth. In a nuclear reactor environment, the bombardment of the materials with energetic neutrons from the fuel fission processes may also affect material properties over time. Since hardness testing is essentially a non-destructive test, it can be applied to components which may, feasibly, continue in service after testing.

The Meyer hardness method is presented as an additional tool to obtain fundamental information from a material as well as a hardness number. The prerequisites for Meyer hardness testing are discussed in terms of test machine and material surface condition requirements. The analysis and interpretation of Meyer hardness data obtained at ambient temperature is described and various empirical correlations are given for the case of a low carbon structural steel, hardened by cold rolling, in various stages of annealing.

The case is presented for including Meyer hardness testing as a part of an integrated component testing programme, if hardness changes occurring in the material are an indication of ageing degradation.

1 INTRODUCTION

The material property of hardness has been used traditionally to determine consistency in the production of components and to test for conformity to specifications. There is, however, another emerging important role of hardness testing as a control on the state of components which have been in service. For the case of safety-related components, it is desirable to monitor their mechanical property state as time at temperature or, in the case of a nuclear reactor pressure vessel, neutron fluence is accumulated during service since thermally activated processes may cause changes in the material's structure and, potentially, its mechanical properties (ageing).

Hardness tests are frequently used in the production of components to control quality aspects and also in basic research into alloying behaviour and thermal stability studies for example. The attractions of a hardness test such as the Brinell (Brinell 1973) or Vickers (Vickers 1976) lie in their fundamental simplicity in concept, the relative ease with which they can be used and the rapidity with which a result can be obtained. The hardness numbers thus obtained give a measure of the material's resistance to the penetration of a hardened ball (Brinell method) or a diamond pyramidal indenter (Vickers method); a low number indicates softness and a high one hardness. An empirical correlation between hardness and tensile strength can also be used to estimate tensile properties of materials when no actual tensile test data exist (Brinell 1976).

The need often arises, however, to assess the toughness properties of aged materials and it is here that the hardness test can be unreliable, depending on the material under investigation. In some cases, the property of hardness can empirically correlate well with toughness (the lower the hardness, the tougher the material), but in others it may indicate softness whilst the material remains relatively brittle due, for example, to grain boundary carbide formation in high alloy steels and decohesion effects present in the grain boundaries of susceptible steel through phosphorus enrichment due to diffusion (non-hardening embrittlement). Hardness tests are essentially measures of the bulk structure property; local hardening at grain boundaries is less likely to be detected in a hardness test. Thus, performing hardness tests alone may have pitfalls with potentially serious consequences.

This work deals with the hardness test (ball penetration) proposed by Meyer (Meyer 1908). In many aspects the Meyer test is similar to the Brinell test but it can yield much more fundamental information about the physical state of a material. A description of the Meyer analysis and hardness test method follows and an example of how it may be used to follow the various annealing stages of a cold rolled steel is given. The data obtained is discussed in terms of applicability to judge the actual physical state of the material. The potential of the method for application to monitor degradation in aged components made from materials which are likely to undergo changes in hardness is indicated.

The principal object of the present work was to investigate the Meyer technique and then to assess its general applicability and sensitivity to detect hardness and/or structural changes in heat treated low carbon steel (model material). With this work, it is hoped to motivate further investigations on service-aged materials using the Meyer method.

2 EXPERIMENTAL PROCEDURES

2.1 *Material*

A cold rolled low carbon structural steel was selected as the model substance to be investigated. The approximate chemical composition was C 0.15, P 0.05, S 0.05 with balance Fe (all weight %). The as-received steel plate was annealed under vacuum at 750°C for 3 hours and furnace cooled before being cold rolled from 20 mm to 10 mm thickness in 5 passes of 2 mm reduction each pass. The plate was annealed again and then finally reduced to 5 mm thickness in 5 passes of 1 mm. This was to ensure a high degree of cold work and hardening in the structure. Specimens in the form of coupons 20 x 20 x 5 mm were obtained by electro-erosion; polishing of the surfaces perpendicular to the direction of rolling (the test surface) was carried out using standard techniques with diamond paste down to 0.5µm finish.

2.2 *Heat treatments*

All heat treatments were performed under vacuum to prevent surface oxidation. Heating rates were relatively rapid due to the small specimens and large furnace; cooling was achieved by quenching the evacuated quartz glass ampoules containing the specimens in water. The cold rolled low carbon steel was subjected to a series of isothermal treatments to develop various levels of recovery, recrystallization and grain growth in the structure, see Figure 1 (a) and Figure 1 (b). Heat treatments are summarised in Table 1 which also includes the Meyer analysis parameters and hardness numbers and other data for ease of comparison.

2.3 *Meyer hardness test*

The Meyer hardness test method consists of slowly applying a loaded hardened steel or titanium carbide ball indenter to the surface of the test specimen, waiting until all deformation has ceased (usually 15 s is sufficient), removing the indenter and measuring the diameter of the

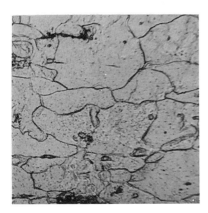

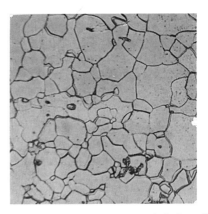

Figure 1 a. Low carbon steel in the 50% cold rolled state. The grains are elongated in the direction of rolling; Meyer hardness (MH) = 209. Etched in 2% nitric acid-methyl alcohol (Nital).

Figure 1 b. The same steel isothermally annealed for 90 min at 575° C; the material has recrystallized and the hardness has decreased to (MH) = 142, corresponding to the fully softened state.

resulting impression (Meyer 1908, Dieter 1961). The method is thus similar to the standard Brinell test (Brinell 1973). However, whereas the Brinell hardness (BH) number is then calculated as the load divided by the spherical contact area, the Meyer hardness (MH) number (or average pressure) is calculated as the load divided by the projected area of the impression. A comparison between the Brinell (BH) and Meyer (MH) hardness number formulas shows this basic difference:

$$BH = P/(\pi D/2[D - \sqrt{(D^2 - d^2)}]) \tag{1}$$

$$MH = 4P/ (\pi d^2) \tag{2}$$

where (P) = the applied load, (D) = the diameter of the spherical indenter and (d) = the diameter of the impression left on the test surface.

The units of hardness are therefore load/area according to equations (1) and (2). The usual units are given as kg/mm^2; the conversion to force/area is obtained by multiplying the load by 9.81 to give Newton (N)/mm^2. It should be noted here that the hardness numbers are given hereafter without the units for convenience and that converting to a force expressed in Newton does not affect the following analyses. The load required for valid Brinell testing is given in the standard procedure (Brinell 1973). For an indenter diameter (D) of 1.25 mm (the present case) a load of 46.875 kg (459.84 N) is needed for steel according to (Brinell 1973):

$$30 = P/D^2 \tag{3}$$

where P = the load (kg) to be used and D = the ball indenter diameter (mm).

The Meyer method requires that testing is carried out in the plastic region of the material's stress-strain curve; in the present case for cold rolled low carbon steel (hard condition) this was found to be at a load of 20 kg (196.2 N). The assumption used is that the deformation under the indenter is mostly elastic until the average pressure exceeds about 2.8 times the material's yield stress (Dieter 1961).

2.4 Meyer's law and analysis

Meyer's law (Meyer 1908) may be expressed as:

287

$$P = K (d/D)^n \qquad (4)$$

where (P) = applied load (kg), (K) = a constant expressing the material's resistance to penetration (kg), (d) = the diameter of the indent (mm), (D) = the diameter of the indenter ball (mm) and (n) = an index related to the strain hardening capacity of the material (Dieter 1961).

For a given material state (i.e. (K) and (n) constant) equation (4) transforms to a linear form when logs are taken:

$$\log P = \log K + n \log (d/D) \qquad (5)$$

where (n) is obtained from the slope of the plot of log (P) as a function of log (d/D) and (K) = the value of (P) when (d/D) = 1.

The physical interpretation of this is the value of load (P) required to press the indenter ball into the material, up to its equator; in the present case giving an indent diameter (d) of 1.25 mm and therefore a value of (d/D) = 1. It is convenient to mention here that (n) and (K) were obtained, in the present investigation, by using a computer program which automatically converted the input values of (P) and (d/D) to logarithms and then performed a least squares fit on the linear plot. The goodness of fit, on 6 pairs of measurements of (P) and (d/D), was always highly significant (r^2 = 0.995 or better for 4 degrees of freedom, >> 1% significance level (Neville 1968)) showing the validity of the Meyer law under the present experimental conditions. The value of (D) was set at the constant value of 1.25 mm.

The value of P in the Meyer hardness (MH) equation (2) may be expressed in terms of the Meyer law, equation (4) to give the following alternative expression for the Meyer hardness (MH):

$$MH = 4 (K(d/D)^n) / \pi d^2 \qquad (6)$$

(See equations (2) and (4) for symbol explanation).

2.5 Meyer test method

A hardness testing machine (indenter ball diameter (D) = 1.25 mm), having several load choices (see below), is essential for performing the Meyer analysis since this requires a series of load (P) vs. indent diameter (d) readings to be obtained (equation (5)). In the present case, this was done by selecting loads of 20 kg, 30 kg, 31.25 kg, 40 kg, 50 kg and 62.5 kg (196.2 N, 294.3 N, 306.6 N, 392.4 N, 490.5 N and 613.1 N, respectively) and measuring the resulting diameter (d). Errors on diameter (d) measurements were estimated to be +/- 5µm maximum. The hardness testing machine was of the oil and dashpot type and no significant errors in load or frictional effects were present. Therefore, the only significant source of error lay, potentially, in the measurement of the indent impression (d) on the test specimen's surface. A well-polished surface is advantageous from this aspect, enabling well-defined indent impressions to be obtained.

The technique employed was to select progressively higher loads and to replace the indenter into the previous impression. In this manner, a series of load (P)-indent diameter (d) measurements were obtained throughout the full plasticity range from a load of 20 kg to 62.5 kg (196.2 N to 613.1 N respectively) giving 6 measurement pairs (d/D vs. P) in total.

3 RESULTS

The results of the Meyer parameter investigations are presented in Table 1. A comparison between the Meyer (MH) and Brinell hardness (BH) numbers (equations (1) and (2)) for the same calculated values of (d), using the Meyer law, at the valid Brinell load of 46.875 kg (force

Table 1: Heat treatments, Meyer parameters, Meyer (MH) and Brinell (BH) hardness.

Temp. °C	Time min.	20(kg) d/D	MH	30(kg) d/D	MH	31.25(kg) d/D	MH	40(kg) d/D	MH	50(kg) d/D	MH	62.5(kg) d/D	MH	K(kg)	n	d* (mm)	MH ᵞ	BH*	Δ MH
Cold rolled	0	0.280	208	0.344	207	0.352	206	0.392	212	0.440	210	0.496	208	259.1	2.014	0.535	209	199	0
550	50	0.288	196	0.344	207	0.352	206	0.400	204	0.448	203	0.504	200	252.4	2.015	0.542	203	192	4
	100	0.296	186	0.352	197	0.360	197	0.408	196	0.456	196	0.512	194	251.7	2.058	0.552	196	185	8
	200	0.304	176	0.360	189	0.368	188	0.416	188	0.464	189	0.520	188	250.9	2.100	0.562	189	178	12
	370	0.320	159	0.384	166	0.392	166	0.440	168	0.488	171	0.544	172	232.5	2.146	0.593	170	159	13
575	10	0.284	202	0.344	207	0.352	206	0.400	204	0.448	203	0.496	207	255.9	2.020	0.539	205	196	5
	45	0.306	174	0.368	181	0.376	180	0.424	181	0.472	183	0.528	183	238.5	2.083	0.573	182	172	9
	60	0.336	144	0.408	147	0.416	147	0.464	151	0.512	155	0.576	153	205.9	2.141	0.626	152	142	9
	90	0.352	132	0.424	136	0.432	136	0.480	141	0.536	142	0.592	145	197.4	2.192	0.649	142	131	13
585	10	0.288	196	0.352	197	0.360	197	0.408	196	0.456	196	0.504	201	246.7	2.021	0.550	198	188	4
	15	0.312	167	0.376	173	0.384	173	0.432	175	0.480	177	0.536	177	233.7	2.105	0.583	176	166	10
	20	0.340	141	0.416	141	0.424	142	0.472	146	0.520	151	0.576	153	206.5	2.182	0.633	149	139	12
600	2	0.280	208	0.336	211	0.344	215	0.384	221	0.432	218	0.488	214	279.0	2.054	0.525	217	208	6
	10	0.344	138	0.416	141	0.424	142	0.472	146	0.528	146	0.584	149	199.0	2.154	0.639	146	136	11
	20	0.352	132	0.424	136	0.432	136	0.480	141	0.536	142	0.592	145	197.4	2.192	0.649	142	131	13
650	0.5	0.276	214	0.336	217	0.344	215	0.388	217	0.436	214	0.488	214	262.7	2.000	0.527	215	204	0
	1	0.280	208	0.344	207	0.352	206	0.392	212	0.440	210	0.496	208	259.1	2.014	0.535	209	199	0
	3	0.312	167	0.376	173	0.384	173	0.432	175	0.480	177	0.536	177	233.7	2.105	0.583	176	166	10
	3.5	0.352	132	0.424	136	0.432	136	0.480	141	0.536	142	0.592	145	197.4	2.192	0.649	142	131	13
700	0.5	0.276	214	0.336	217	0.344	215	0.388	217	0.436	214	0.488	214	262.7	2.000	0.527	215	204	0
	2.0	0.352	132	0.424	136	0.432	136	0.480	141	0.536	142	0.592	145	197.4	2.192	0.649	142	131	13

Notes:

(d/D) is ratio of indent diameter (d) to indenter diameter (D = 1.25 mm), (MH) is the Meyer hardness number calculated from equation (2) for the load (kg) shown. (K) and (n) are the Meyer constant (kg) and exponent respectively from the Meyer law:
$P = K(d/D)^n$ (see equation (4)).

The values of (d*) were calculated from the Meyer law parameters and the theoretical load required for valid Brinell hardness testing of 46.875 kg (equation (3)) The values of (d*) thus obtained were then used in the Meyer and Brinell hardness number formulas (equations (1) and (2)) to obtain the Meyer hardness number (MH*) and the standard Brinell hardness number (BH*) at the valid load for standard Brinell testing (equation (3)) to facilitate comparison between the two; see also Figure 2. The difference in Meyer hardness number (ΔMH) is obtained from the numerical difference in (MH) calculated from the highest and lowest loads. The Meyer hardness (MH) number is dependent on the applied load only for the case of material capable of strain hardening.

of 459.84 N) is shown in Figure 2. The relationship between the Meyer constant (K) and the Meyer hardness (MH) number, calculated at a load of 46.875 kg is shown in Figure 3. The difference in Meyer hardness number between the highest and lowest loads applied per test (material condition) (ΔMH) is plotted as a function of the Meyer exponent (n) in Figure 4.

4 DISCUSSION

It can be seen from Figure 2 that the Meyer hardness number is comparable, as a measure of hardness, with the Brinell number. The Meyer hardness number is higher than the Brinell number for a given value of indent diameter (d) at a given load of 46.875 kg because the former is calculated from the projected area (equation (2)) as opposed to the Brinell number which is calculated using the spherical surface area, which is larger (equation (1)).

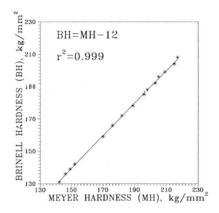

Figure 2. The Brinell (BH) (equation (1)) and Meyer (MH) (equation (2)) hardness values compared at the same value of indent diameter (d) and the valid load of 46.875 kg for the Brinell test. The values of (d) for this load were calculated using the Meyer law for the corresponding material condition. Note that the complete range of material conditions from cold rolled (hard) to fully annealed (soft) are present in the data shown.

For annealed and softened steel in various stages of recovery and recrystallization, the value of the Meyer hardness (MH) number is dependent on the load used to obtain it; the Meyer exponent (n) is greater than 2.0, being typically around 2.19 for this type of material in the fully soft condition (Table 1). For hard (cold rolled) material, which is heavily reduced in further strain hardening capacity, the value of the hardness number is sensibly constant and relatively high and not dependent on the load applied to obtain it (at least in the range of loads applied here), Table 1. The Meyer exponent (n), for the perfect plastic state, is ideally equal to 2.0 (Dieter 1961); (Table 1). In the present case, the value of (n) was slightly higher, at 2.014 for the cold rolled state (Table 1). This shows that, although the material was heavily cold-worked, it still had a slight capacity left for strain hardening and was not in a "perfect plastic" state. The material state was, however, quite near to the "perfect plastic" one. Accordingly, the material can be identified as being in a fully softened, quarter hard, half hard, nearly full hard or fully hardened (cold rolled) state not only by the Meyer hardness (MH) number (low or high) but also by the Meyer exponent (n) (high or low) values respectively. It is convenient to note here that (MH), (K) and (n) have also been shown to be sensitive in detecting material degradation in fast neutron irradiated low alloy ferritic materials (Tipping 1995). Additionally, the Meyer constant, (K), defined as the load required to press the spherical indenter up to its equator (i.e. when d/D = 1) into the specimen, is also highly associated with the Meyer hardness number (MH); the softer the material, the lower the value of (K); see Figure 3. Inspection of Table 1 shows that the indent diameter (d) is dependent on both (K) and (n) for a given material state: keeping the load (P) constant at 46.875 kg, (d) is smaller at high values of (K) and higher at higher values of (n). This is physically sensible. An understandable concept of hardness is therefore provided by (K) using this interpretation. The value of (K) will depend on the diameter of indenter ball (D) used on a material in a given physical state and it is therefore important to provide this information on (D) when (K) values are compared (Meyer 1908, Dieter 1961). The (K) values are therefore not a fundamental material property. On the contrary, the value of (n) has been found to be nearly independent of the indenter diameter (D) and as such, reflects some fundamental material property related to the strain-hardening capacity (Dieter 1961). It should be noted that the association between the Meyer constant (K) and the Meyer hardness (MH) number for the present conditions using a 1.25 mm diameter indenter ball (D) is valid only in the full plasticity range of values of load for the material; extrapolation of the fitted equation to lower values will therefore have no physical sense due to the violation of the requirement for full plasticity conditions. Low loads will yield smaller than ideal values of (d) due to elastic recovery effects.

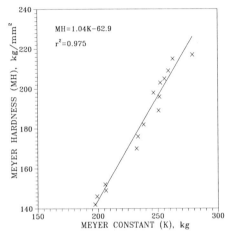

Figure 3. Meyer hardness (MH) number (calculated at load = 46.875 kg, equation (2) and using calculated value of (d) at this load via equation (4)) as a function of the Meyer constant, (K). The data (Table 1) indicates a good fit between the two parameters and reflects the physical meaning of (K) as a measure of the material's resistance to penetration by a loaded ball indenter. The data covers many physical states of the material (from cold rolled (hard) to fully recrystallized (soft)); see also Figures 1a and 1b and 2.

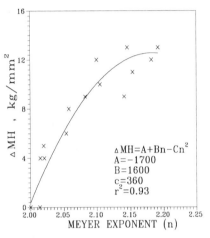

Figure 4. The difference in Meyer hardness (ΔMH) for each material condition (obtained by subtracting the hardness values obtained from the highest load (62.5 kg) and the lowest load (20 kg)) as a function of the Meyer exponent (n). The data indicates the physical meaning of (n) as indicating the presence of strain hardening capacity in the material. At (n) $\cong$ 2.0, the value of (ΔMH) is virtually zero indicating a fully cold rolled condition and highly reduced strain hardening capacity. In this condition the Meyer hardness (MH) number is sensibly constant and independent of the load applied to obtain it (Table 1).

Inspection of Table 1 reveals that the Meyer hardness (MH) and constant (K) of some specimens annealed at relatively high temperatures for short times actually increased slightly in hardness before softening at longer times. Also, the Meyer exponent (n) increased to 2.054 in one case. Specifically, the data indicating this behaviour comes from the following isothermal annealing conditions: 600°C for 2 minutes (MH = 217, n = 2.054, K = 279.0 kg), 650°C for 30 s (MH = 215, n = 2.000, K = 262.7 kg) and 700°C for 30 s (MH = 215, n = 2.000, K = 262.7 kg). It should be noted that due to the very short times involved here, heating up and quenching effects may be affecting the actual isothermal conditions strived for.

The cold rolled condition hardness was (MH) = 209, with (n) = 2.014 and (K) = 259.1 kg, Table 1. The above effect of slight hardness increase is suspected as being real since the hardness testing procedure was based on multiple tests to optimise the values of (d) and to obtain accurate values of hardness and best fit values of (K) and (n) parameters. The observed slight hardening effect may be due to an age-hardening mechanism in the very early stages of the recovery process; an effect of carbon atmospheres diffusing to better fully decorate the dislocation tangles could, feasibly, lead to an additional slight hardening contribution. The effect may also be related to the blue-brittleness phenomenon in some way. Later on, when more time has been accumulated at temperature, the dominance of the recrystallization and grain growth processes becomes evident as general softening takes place and carbon atom atmospheres boil off as the dislocations straighten out and reduce in density. The increase in (n) to 2.054 for the isothermal condition of 600°C for 2 minutes may be an indication of the rapid restoration of strain hardening capacity in the early recovery stage of the annealing process.

A specific study using transmission electron microscopy and perhaps, internal friction experiments etc., which are out of the scope of the present paper, would be necessary to investigate the above assumptions. However, the indications here are that the Meyer hardness test and analyses are quite sensitive and capable of yielding more basic information about a material's physical state than other hardness tests.

The general tendency for (n) to increase as (K) and Meyer hardness (MH) decrease is evident for many material conditions (degree of annealing), see Table 1. This also indicates that the material in the softened condition is capable of undergoing strain hardening.

For material in the softened or even partially recrystallized state, the value of Meyer hardness number depends on the load applied to obtain it within the full plasticity range. In the present case, the lowest load applied was 20 kg (96.2 N) (start of validity for Meyer hardness) and the highest practical load (test machine limitations) was 62.5 kg (613.13 N). The hardness numbers as a function of the load used to obtain them (equation (2)) describe a smooth curve (not linear) within the plasticity range, between the lower and upper values of applied load. This can be seen by inspecting the relevant data in Table 1.

If the Meyer hardness (MH) numbers are calculated (equation (2)) only for the two extreme experimental loads of 20 kg (96.2 N) and 62.5 kg (613.13 N) and the difference taken (ΔMH) and plotted as a function of the corresponding value of Meyer exponent (n), an association is found; see Table 1 and Figure 4. The data indicate a curve which tends towards a saturation value at the maximum value of n = 2.192 and ΔMH = 13. The apparent rate of increase of (ΔMH), as annealing (softening) occurs with progressive isothermal annealing condition (Table 1), slows down as the saturation levels are approached, see Figure 4. The main stages of annealing of deformed metals are 1) recovery (change in the distribution and density of defects introduced by the cold work), 2) recrystallization (the distorted grains of the metal are replaced by essentially stress-free ones) and 3) grain growth (smaller grains disappear to supply material to larger ones). The physical interpretation of the behaviour observed in the present work may lie in the initial rapidity of the recovery stage due to the high level (50%) of cold work present (high stored energy state which the material seeks to lower with thermally activated processes). Later on in the annealing stages, when stored energy has been dissipated, then also the driving force for softening and regeneration of the work-hardening capacity is less and the rate of both (n) and (ΔMH) recovery is lower. The difference in Meyer hardness number calculated for the two extreme loads used here is an expression of the material's capacity for resisting deformation; the Meyer exponent is also related to the strain-hardening capacity (Dieter 1961). The (n) and (ΔMH) values are reduced progressively as the material approaches a cold rolled and highly strain hardened state. The value of (ΔMH) approaches zero when (n) approaches the theoretically lowest value of 2.0 (Dieter 1961). In the heavily cold rolled state the material's dislocation density is likely to be very high and a lot of dislocation entanglement is expected; the capacity for further significant strain hardening is therefore low since the dislocations are already highly interacted and cannot interfere with each other's movement anymore. The exponent (n) approaches the value of 2.0, (ΔMH) approaches zero and the Meyer hardness number is constant (and high) and independent of the load applied to obtain it (not much strain-hardening effect present). This general behaviour has been noted previously (Meyer 1908, Dieter 1961, Tipping 1995).

It is convenient to mention here that the Meyer exponent (n) is also related to the strain hardening coefficient in the exponential equation for the true stress-true strain tensile curve as given by:

$$\sigma = C\varepsilon^{n'} \tag{7}$$

where (σ) = the true flow stress, (C) = a constant, (ε) = the true strain and (n') = an exponent related to strain hardening in tension.

The exponent (n) in Meyer's law is approximately equal to the tensile strain-hardening coefficient (n') plus 2 (Dieter 1961). Therefore, at least theoretically, it is possible to obtain an estimate of the tensile strain-hardening exponent by performing the Meyer analysis to derive (n) and then subtracting 2 to obtain (n'). Ideally, when (n) = 2, then (n') = 0 and once a minimum value of tensile stress has been exceeded, the material will flow plastically at a constant flow stress: this defines the perfectly plastic, rigid material and the absence of strain hardening. In real engineering materials such as low alloy carbon steel, there is an elastic part to the stress-strain curve before yielding and plastic flow occurs; strain hardening is then seen as an increase in stress for each strain increment: strain hardening is present.

The true strain (ε), is proportional to the ratio of (d/D) (Dieter 1961, Tabor 1951):

$$\varepsilon = 0.2 \, (d/D) \tag{8}$$

where (ε) = the true strain, and (d) = the indent diameter and (D) = the spherical indenter diameter. Thus, if the Meyer technique is used under conditions of full plasticity (i.e. minimum load, (P) ≥ 2.8 times the yield stress) and (d/D) vary as a function of load (P), which will be the case for material capable of strain-hardening, then it is also possible to approximate the tensile flow behaviour (Tabor 1951).

5 CONCLUSIONS

The Meyer analysis and hardness testing method has been used to characterise the response of a cold rolled low carbon structural steel to recovery, recrystallization and annealing (softening). Based on the results, it can be seen that the room temperature Meyer hardness (MH) and constant (K) and exponent (n) can be used to detect and interpret changes which have occurred in the original material state.

The Meyer method requires a good surface condition of the test piece and a standardised machine to minimise errors in measured indent diameter (d); all Meyer parameters are very sensitive to errors in (d).

The Meyer (MH) and Brinell (BH) hardness numbers are linearly correlated in the range of loads used; at a given load (46.875 kg), valid for the Brinell test under these conditions and values of indent diameter (d) calculated via the corresponding Meyer's law parameters for the given material state, the Brinell number is less than the Meyer one. This is due to the effect of the larger spherical contact area used in the Brinell hardness formula compared to the projected area in the Meyer hardness formula (equations (1) and (2)).

For a given material condition, the indent diameter (d), at a given load, is dependent on both the strain-hardening related index (n) and the Meyer constant (K).

The Meyer hardness (MH) number and constant (K) both decreased as annealing took place under progressive isothermal treatments favouring recovery and recrystallisation of the cold rolled structure. The Meyer index (n), related to the strain-hardening capacity, and the difference in Meyer hardness (ΔMH) obtained from the difference in Meyer hardness (MH) number using the highest and lowest loads in the plasticity range, both increased as softening took place.

The Meyer hardness test yields information about a material's strain-hardening capacity and resistance to penetration. Correlations exist between the Meyer parameters (K) and (n) with the hardness (MH) number and difference in hardness number obtained with low and high loads (ΔMH) respectively.

Cold rolled material is hard and has a reduced capacity for strain-hardening; its condition was near to a "perfect plastic" one, but slight strain hardening still occurred. Material in the soft (annealed) condition has a higher capacity for strain hardening.

Tensile flow characteristics of the material may be estimated using the relationships between the Meyer hardness (MH) and the true flow stress (σ) and the ratio of indent diameter to indenter diameter (d/D) and the true strain (ε).

The main emphasis here has been to develop experimental techniques and to examine the various associations between the hardness numbers and Meyer analysis parameters with a view to assessing the material's physical state. The data and analyses indicate the usefulness and potential of the Meyer method to detect changes in hardness-related properties in a cold rolled and annealed low carbon steel.

DISCLAIMER:

The ideas and interpretations expressed herein are those of the authors and these need not agree with those of their respective affiliations.

REFERENCES

Brinell, 1973: Brinell hardness test: German Industrial Standard DIN 50351.
Brinell, 1976: Hardness comparison tables and tensile strength: German Industrial Standard DIN 50150.
Dieter, G. 1961. *Mechanical Metallurgy*, New York: McGraw-Hill.
Meyer, E. 1908. Untersuchungen über Härteprufung und Härte. *Zeitschrift des Vereines Deutscher Ingenieure*, 52: 17 645-654.
Neville, A. & Kennedy, J. 1968. *Basic Statistical Methods for Engineers and Scientists*, London: Intertext Books.
Tabor, D. 1951. *The Hardness of Metals*, New York: Oxford University Press.
Tipping, Ph. & Cripps, R. 1995. Neutron irradiation sensitivities of mock-up ASTM A 508 Class 2 PV base plate and automatically deposited weld material: a comparative study using Meyer's hardness, *Int. J. Pres. Ves. & Piping,* 61: 77-86
Vickers, 1976: Vickers hardness test: German Industrial Standard DIN 50133.

Non-metallic materials

Ageing of Materials and Methods for the Assessment of Lifetimes of Engineering Plant, Penny (ed.)
© *1997 Balkema, Rotterdam, ISBN 90 5410 874 6*

Lifetime, toughness and reliability of engineering thermoplastics

A.Chudnovsky, D.Baron & Y.Shulkin
Department of Civil & Materials Engineering, University of Illinois at Chicago, Ill., USA

ABSTRACT: The existing empirical methods of lifetime prediction for engineering thermoplastics, based on extrapolations of the data obtained under high temperatures and stresses to the range of low temperatures and stresses, are reviewed. A new method is proposed as a combination of mathematical modeling of slow crack growth in polymers and experimental determination of the material parameters for the model. The experiments are conducted on smooth specimens and deal with short-term observations. The descriptions of crack behavior and prediction of lifetime resulting from this approach are in good agreement with observations for various engineering thermoplastics. For a particular material, specimen geometry, applied stress level and temperature, the method establishes the dependence of lifetime on initial crack size. This size is a randomly distributed quantity. Its probability density is evaluated in a quality control test. Then, a standard probability treatment produces the probability density for lifetime. This makes it possible to calculate the reliability function of the structural component. Thus, the method allows the manufacturer of polymeric structures to recommend a replacement interval (dependable lifetime) for any value of reliability desired.

1 INTRODUCTION

The failure of engineering thermoplastics typically occurs as a result of either a shear rupture or a sudden onset of instability in a previously slowly growing crack. In both cases the time to failure depends greatly on temperature. Extensive data regarding the effect of temperature upon the relation between the lifetime and applied stress for polymeric materials was reported in [Williams 1987, Lu & Brown 1986, 1990, Huang & Brown 1990]. It was demonstrated [Lu & Brown 1990, Popelar, Popelar & Kenner 1990, Popelar, Kenner & Wooster 1991] that special treatments of the experimental results obtained in short-term fracture tests (at high temperatures) under creep conditions, allow the prediction of the long-term lifetime (at room temperature).

A different approach to lifetime prediction is the theory of crack growth in viscoelastic media [Schapery 1975] (see also references contained therein). Application of the theory to polymers reveals the following discrepancies: (i) the theory predicts a smooth crack propagation under creep, whereas as has been observed, the process proceeds in a stepwise manner [Hertzberg & Manson 1980, Lu, Qian & Brown 1991]; (ii) the stress dependence of lifetime at fixed temperature established theoretically is similar to that commonly observed, and can be approximately expressed as a power law, however, the predicted range of the exponent values differ appreciably from that obtained experimentally [Lu & Brown 1986, 1990, Huang & Brown 1990].

The method of lifetime prediction for polymers under creep proposed below is based on a kinetic model for slow crack grow [Kadota & Chudnovsky 1991, Stojimirovic & Chudnovsky 1992, Stojimirovic, Kadota & Chudnovsky 1992, Kadota & Chudnovsky 1992, Chudnovsky & Kadota 1993, Kadota, Chum & Chudnovsky 1993]. In this model a slow growing crack and the process zone (PZ) which always precedes it are treated as a closely coupled system called the crack layer (CL). The CL model allows the determination of the lifetime for a given structural element under prescribed service conditions, and predicts a lifetime-stress relation identical to the one found experimentally.

A great need for methods to predict expected lifetimes is suggested by the increasing usage of plastics in engineering structures, and by the rising cost for potential product liability. Classical strength theories have been created for elastic brittle materials which account for strength reduction due to the presence of flaws. However, recent studies of field failures of plastics show that they are caused by quasi-brittle cracks grows under low loads [Sehanobish, Moet & Petro 1985, Stivala, Patel, Chudnovsky, Kim & Zhou 1994]. The largest defect in a giving stress state plays the role of a trigger for the process of slow crack growth, leading to brittle failure. The methodology proposed in the present paper is based on the statistical analysis of a defect population together with the computation of the lifetime, to determine the part reliability. This approach allows the manufacturer of polymeric structures to recommend a replacement interval (dependable lifetime) for any value of reliability desired.

2 EMPIRICAL METHODS FOR LIFETIME PREDICTION

The experimental dependence of time to failure t_f on applied stress σ and temperature T is schematically depicted in Figure 1. Within both modes of failure, ductile and brittle, the time-stress relation at fixed temperature can be approximately expressed as follows:

$$\log t_f = A - \alpha \log \sigma \qquad (1)$$

were the symbols t_f and σ represent dimensionless quantities corresponding to the lifetime in minutes and stress in MPa, respectively. The coefficient α sometimes depends on temperature. For example, for a polymer such as polyethylene (PE) [Lu & Brown 1990] α changes noticeably increasing from 3.0 to 4.8 when the temperature increasing from 42 to 80°C. The values of α for ductile and brittle modes differ greatly (10-15 times according to [Lu & Brown 1990]). The stress corresponding to the intersection of the two branches, ductile and brittle, is called the critical stress and denoted by σ_c.

If the coefficient α is independent of temperature, the ductile and brittle branches from two sets of parallel lines. Lines for different temperatures can be combined into one line by shifts directed along the arrow shown in Figure 1. Following [Lu & Brown 1990], one can present the horizontal component of the shifts in the form:

$$\Delta h = \log e \frac{Q}{R}\left(\frac{1}{T_2} - \frac{1}{T_1}\right) \qquad (2)$$

were Q is the activation energy of the process (in [Lu & Brown 1990], the value $Q \cong 86$ kJ/mol is accepted), and R is the universal gas constant ($R = 8.31$ J/K · mol). The vertical shift is expressed in terms of the temperature dependence of the critical stress σ_c:

$$\Delta v = c\left(\frac{1}{T_2} - \frac{1}{T_1}\right) \qquad (3)$$

where, according to [Lu & Brown], the coefficient can be evaluated as $c = 0.42 \times 10^3$ K. Shifts (2) and (3) permit the reduction of all lines at different temperatures to a single one for room temperature and obtain as a result, a master curve of "$\log t_f$ vs. $\log (\sigma / \sigma_{dr})$".

Two different shifts [Popelar, Popelar & Kenner 1990, Popelar, Kenner & Wooster 1991],

$$\Delta h = \log e \times 0.1090(T_1 - T_2) \qquad (4)$$
$$\Delta v = \log e \times 0.0116(T_1 - T_2) \qquad (5)$$

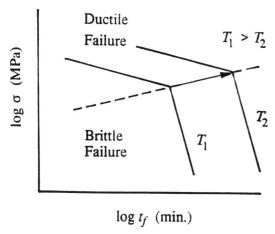

Figure 1. Schematic representation of experimental time-stress relations.

are proposed as universal for the medium density and high density polymers that are commonly used in engineering structures. To determine how (2) and (3) are related to (4) and (5), note first that within the temperature range of 293-353°K (20-80°C)

$$\frac{1}{T_2} - \frac{1}{T_1} \cong 0.96 \times 10^{-5}(T_1 - T_2)$$

with a maximum error of about 15%. If in (2) one accepts $Q = 94$ kJ/mol, this shift can be equated with (4) (with the same maximum error). Similarly, if $c = 0.53 \times 10^3$ K in shift (3), it can be expressed in the form (5). Therefore, the above two sets of shifts are essentially the same. Quantitatively, they should be considered as certain, empirically evaluated quantities, that reflect on average the relationship between time to failure and stress, at various temperatures. Despite a good average agreement, in some cases the lifetime predicted by [Popelar, Popelar & Kenner 1990, Popelar, Kenner & Wooster 1991] may noticeably deviate from that measured directly (10 times and even more). The most important drawback of these methods is common to all empirical approaches: the limitation of the method applicability, and the error resulting from its application, are a priori unknown.

The following considerations are concerned with the brittle mode of failure. From a practical point of view, it is the most important case, since the polymers in engineering applications serve mostly under relatively low stresses.

3 THE CRACK LAYER KINETIC MODEL

Since the PZ is commonly observed in the shape of a thin strip, the CL can be characterized by two parameters - the crack and PZ lengths, l and L (Figure 2). The bulk material surrounding the CL is the original material, while the PZ consists of material identical to that formed in cold drawing (necking) under tension. This material can be treated as a collection of fibers and can be modeled as an unidirectional continuum. The stresses acting between the bulk material and PZ are distributed uniformly along their boundary. These stresses are controlled by the drawing process, and are equal to the drawing stress σ_{dr}.

The driving forces for the crack and PZ advances are determined as follows [Kadota & Chudnovsky 1992, Chudnovsky & Kadota 1993, Kadota, Chum & Chudnovsky 1993]:

$$X_{CR} = -\frac{\partial G}{\partial l} \ , \ X_{PZ} = -\frac{\partial G}{\partial L}$$

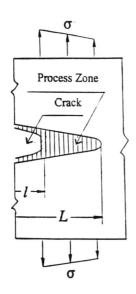

Figure 2. Schematic presentation of crack layer.

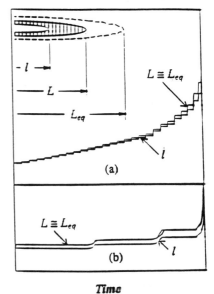

Figure 3. Discontinuous crack layer growth for (a) low stress and (b) high stress.

where G is the Gibbs potential of the system. Since the crack and PZ sizes cannot decrease, $X_{CR} = 0$ and $X_{PZ} = 0$ if the corresponding derivatives of G are not positive. The driving forces depend on properties of the original and drawn materials, including the energy absorbed within the PZ during its formation, as well as on specimen geometry and loading conditions.

The process of CL propagation is governed by equations [Kadota & Chudnovsky 1992, Chudnovsky & Kadota 1993, Kadota, Chum & Chudnovsky 1993]

$$\dot{l} = k_1 X_{CR} \, , \quad \dot{L} = k_2 X_{PZ} \qquad (6)$$

with the initial conditions $l = l_0$ and $L = L_0$ at $t = 0$. If $X_{CR} = 0$ and consequently, the crack is in a steady state, then the equilibrium size of the PZ, L_{eq}, is determined by the equation $X_{PZ} = 0$.

The drawn material within the PZ experiences degradation leading to a decrease of fracture energy. This process is described by equation

$$\gamma = \gamma_0 \left(1 - \frac{t}{t_r} \right) \qquad (7)$$

in which γ and γ_0 are the specific fracture energy at the current and initial moments, and t_r is the time to rupture of fibers under stress σ_{dr}. According to the thermoactivation theory of fracture [Zhurkov 1965], lifetime t_r is considered to be a function of temperature T and stress σ_{dr}:

$$t_r = t_0 \exp \frac{Q_0 - \chi \sigma_{dr}}{RT} \qquad (8)$$

Here t_0, Q_0 and χ are the characteristic time, activation energy (AE) and coefficient of stress induced reduction of AE for the drawn material, respectively.

Kinetic equation (6) together with equations (7) and (8) of the PZ material degradation are strongly nonlinear. The numerical solution of these equations (the fourth order Runge-Kutta

300

method is used) results in the discontinuous (stepwise) evolution of the CL (Figure 3). As seen, the higher the level of remote stress σ, the more pronounced the discontinuity of the slow crack growth is. These predictions of the model agree well with experimental data [Hertzberg & Manson 1980, Lu, Qian & Brown 1991]: the quasi-brittle crack in polymers grows discontinuously under both constant and cyclic loading.

4 TIME-STRESS-TEMPERATURE RELATION

A characteristic length scale in the kinetic model is established by expression

$$l_* = \frac{2E\gamma_0}{\sigma_{dr}^2} \tag{9}$$

where E stands for Young's modulus of the original material. Two other material characteristics introduced in the previous section, the drawing stress σ_{dr} and the rupture time t_r of the drawn material, are used as scales of stresses and time, respectively. One more physical parameter of the kinetic model is the energy required for the PZ formation; its density is denoted by γ_{dr}. The kinetic model explicitly accounts for the system geometry (sizes of the plate and initial length of the notch), as well as the loading conditions (simple or eccentric tension, pure or three point bending, etc.). For simplicity, all further considerations are confined to simple tension of a SEN specimen.

The following dimensionless variables are introduced:

$$\tau_f = \frac{t_f}{t_r}, \quad \bar{\sigma} = \frac{\sigma}{\sigma_{dr}}, \quad \eta = \left\{ \frac{l_0}{l_*}, \frac{W}{l_*} \right\}$$

where W is the plate width. If the material properties, specimen geometry and applied stress are prescribed, a computer simulation of slow CL growth can be constructed by means of the numerical solution of equations (6)-(8). Such numerical experiments lead to the following approximate relation between the lifetime and applied stress:

$$\log \tau_f = B - \beta \log \bar{\sigma}, \tag{10}$$

that valid for relatively low applied stresses (approximately for $\bar{\sigma} \leq 0.5$). Our analysis shows that quantities B and β only slightly depend on the transformation (drawing) energy γ_{dr}, and can be considered as functions of the characteristic length and specimen geometry only. It is important to emphasize that both the experimental and theoretical relations, (1) and (10) respectively, establish a power dependence of lifetime on applied stress; and the range of exponent β, from 2 to 5, predicted by the kinetic model, coincides with the range of exponent α observed for various kinds of engineering thermoplastics [Lu & Brown 1986, 1990, Huang & Brown 1990].

Using (8), one can express (10) in the form

$$t_f = t_0 \left(\frac{\sigma}{\sigma_{dr}} \right)^{-\beta} \exp\left(\frac{Q_0 - \chi\sigma_{dr}}{RT} + b \right) \tag{11}$$

where $b = \beta / \log e$. Here quantities b and β are the kinetic model parameters reflecting properties of the material and specimen geometry, while t_0, Q_0 and χ are the properties of the original and drawn materials only. Equation (11) is the time-stress-temperature relation predicted by the CL model.

If for a certain material and specimen geometry, the parameters b and β are temperature independent, the transition from one temperature to another can be obtained by means of a horizontal shift like (2) with $Q = Q_0$, and by the vertical shift

$$\Delta v = \beta \log\left(\frac{\sigma_{dr}(T_2)}{\sigma_{dr}(T_1)}\right) - \log e \frac{\chi\sigma_{dr}}{R}\left(\frac{1}{T_2} - \frac{1}{T_1}\right) \tag{12}$$

The two material constants in (11), i.e., characteristic time t_0 and coefficient χ are found from data [Lu & Brown 1990] for temperatures 50 and 70^{0}C: $t_0 = 0.64 \times 10^{-11}$ min., $\chi = 0.63 \times 10^{-3}$ m^3/mol.

The data of the direct measurements [Lu & Brown 1990] (points), and the theoretical predictions of equations (6)-(8) (lines), are combined in Figure 4. The dashed lines correspond to the adjustment temperatures 50 and 70^{0}C. The theoretical results agree with the experimental results (within scatter of data), not only at 60^{0}C (the temperature of interpolation), but also at 42 and 80^{0}C (the temperatures of extrapolation). Experimental data for temperature 24^{0}C do not exist. Evidently, the CL kinetic model correctly describes the stress-lifetime relationships for brittle fracture at various temperatures.

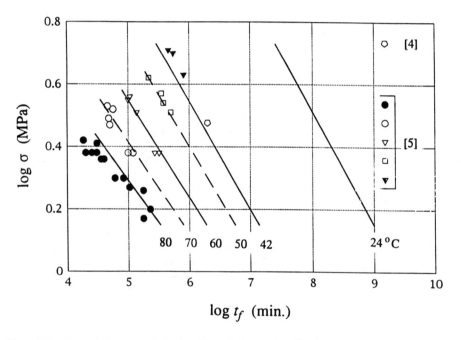

Figure 4. Experimental observations (points) and theoretical predictions (lines).

5 MEASURE OF TOUGHNESS

Toughness is understood as a measure of material resistance to crack initiation and/or crack extension. Various parameters are widely used to characterize toughness. Here we analyze the J-R curve and J_{IC} approaches via computer simulation of crack initiation and growth using Crack Layer model. Figure 5 shows the numerical simulation of the crack and the PZ evolution vs. displacement Δ (the lower part of the Figure 5) and the resulting remote stress σ_∞ vs. displacement Δ in a displacement control test (the upper part of the Figure 5). The J_l as a

302

function of crack extension δl for three different displacement rates Δ is shown in Figure 6 (J_I is computed as $2U(t)\{B[W-1(t)]\}^{-1}$, where U(t) is equal to the area under σ vs. Δ curve at the time "t", B is the specimen thickness and l(t) is the crack length at time "t", according to ASTM E813-81). The parameter J_{IC} is determined as the value of J_I at the intersection of the curve with the blunting line described by equation $J_I = 2\sigma_{dr}\delta l$ with drawing stress σ_{dr} employed instead of yield stress σ_y, since the later has a strong rate dependency. The above computer simulation of rate dependent J_{IC} and J-R curve closely resemble that observed experimentally. It also suggests that there is no reason to expect a constant value of parameters like J_{IC} or K_{IC} to represent material toughness. We propose the criteria of crack instability with a stationary PZ as the crack initiation and the global CL instability as the criterion of catastrophic failure instead of conventional Fracture Toughness parameters.

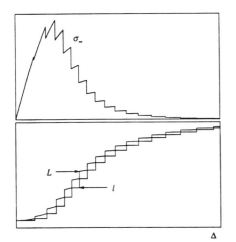
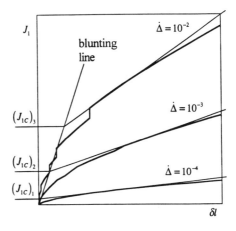

Figure 5. Typical remote stress curve and corresponding CL growth curves, all plotted vs displacement.

Figure 6. J_{IC} calculated according to ASTM E813-81 for three values of displacement rate.

6 RELIABILITY ANALYSIS

The analysis is based on the following assumptions:
(i) The sizes and locations of the defects are random and cause local stress concentrations.
(ii) The stress concentration due to the largest defect results in PZ formation and crack initiation within the PZ.
(iii) The deterministic equations of the CL model are adequate for description of the subsequent slow crack and PZ growth in polymers leading to brittle failure.

Following assumption (ii), the statistics of extremes for a finite interval is taken as the most appropriate to characterize the size distribution of the defects that cause failure.

Destructive tests are the most desirable to obtain the relevant defect population. Fractographic analysis of the fracture surfaces result in identification of the statistics of size and location of the critical defects. For example, in the fatigue of many plastic components, the failure initiation site is generally observed to occur at the surface. Statistics of extremes is usually employed to characterize "the weakest link" from which the failure process begins [Weibull 1951]. Three well known types of the extremes functions exist for both minimal and maximal values [Gumbel 1958]. Recently, these have been generalized into one cumulative distribution function for a finite interval [Kunin 1991]. For maximum size of the initial defects, l_0, the distribution is written as

$$F(l_0) = \left\{ \begin{array}{c} 0: l_0 < l_{min} \\ \exp\left[-\alpha\left(\dfrac{l_{max} - l_0}{l_0 - l_{min}}\right)^{\beta}\right] : l_{min} \le l_0 \le l_{max} \\ 1: l_{max} < l_0 \end{array} \right\}$$ (13)

Here l_{min} and l_{max} are the smallest and largest sizes of the defects for the material in question, and α and β are parameters of the distribution, similar to the scale and shape parameters in the classical Weibull distribution. Differentiation of this function with respect to l_0 gives the probability density function $f(l_0)$. Figure 7 displays the theoretical approximation of the histogram of the critical defect size l_0 obtained from the fractographic analysis of over 50 fracture surfaces in simple tension (the data was kindly provided by Drs. C.P. Bosnyak and K. Sehanobish of Dow Chemical Co., USA).

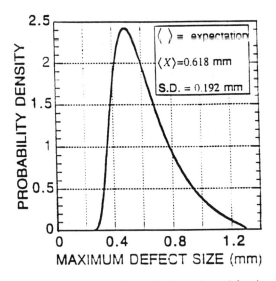

Figure 7. Modeling probability density for maximum defect size.

The total lifetime t_f consists of the time interval required until crack initiation plus the time during which the crack growth from its initial length up to the length for global instability of the process. The crack initiation process in the CL model is defined as the first formation of the PZ, followed by degradation of the PZ material leading to the crack advance. In the case of simple tension, numerical simulations of slow CL growth at moderate stress levels ($\sigma_\infty / \sigma_{dr} \le 0.5$) lead to a relationship between the time to failure t_f and initial defect size, t_f vs. l_0, which can approximately be presented in the form

$$t_f = A + B\log(l_0)$$ (14)

Here coefficients A and B depend on properties of the original and transformed materials, the specimen geometry, applied stress, and temperature. In general, theses coefficients should be regarded as random quantities due to the variations in the factors listed above. However, in the present analysis the randomness of A and B is not considered, since the objective is to outline a methodology for evaluation of reliability associated with the critical defect size. Therefore,

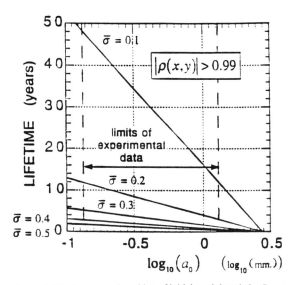

Figure 8. Lifetime versus logarithm of initial crack length for five stress levels.

below lifetime t_f is treated as a deterministic function of one random variable l_0. The dependencies calculated for single-edge notched specimens of polymeric material under various applied stresses at room temperature are shown in Figure 8.

Since t_f is a function of the random initial defect size l_0,

$$t_f = \phi(l_0),\tag{15}$$

given the probability density function $f(l_0)$, a standard technique of probability theory can be employed to compute the probability density function for lifetime:

$$g(t_f) = f\left[\phi^{-1}(t_f)\right]\left|\frac{d\phi^{-1}(t_f)}{dt_f}\right|\tag{16}$$

The reliability function, which provides the probability that a part is still functioning at any chosen time t, depends on the stress level, and is conventionally defined as

$$R(t,\sigma) = 1 - \int_{t_f=0}^{t_f=t} g_\sigma(t_f)dt_f\tag{17}$$

Here g_σ indicates the lifetime probability density (16) parametrized by the applied stress σ_∞. Figure 9 shows a graph of the lifetime versus applied stress for various assigned values of the reliability. A simpler characteristic, the mean time to failure $< t_f >$ may be adequate for ranking materials and testing of a design. This is shown versus remote stress in Figure 10.

7 CONCLUSION

A simple method of lifetime evaluation and material resistance to cracking is presented above. It implies an experimental evaluation of relevant material properties such as elastic, yield, drawing and statistics of defects population resulting from manufacturing conditions as well as from chemical degradation and aging. The above information combined with the CL model presented in the paper allows one to estimate major parameters of long-term brittle fracture.

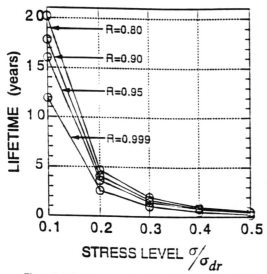

Figure 9. Lifetime versus stress level for four values of reliability.

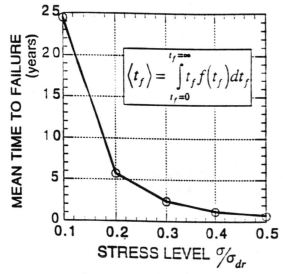

Figure 10. Mean lifetimes for five stress levels.

REFERENCES

Chudnovsky, A. & K. Kadota 1993. *Proceedings PACAM III*: 419.

Hertzberg, R.W. & J.A. Manson 1980. *Fatigue of engineering plastics*. Academic Press.

Huang, Y.L. & N. Brown 1990. *J. Polymer Sci.* 28: 2007.

Gumbel, E. 1958. *Statistics of extremes*. New York: Columbia University Press.

Kadota K. & A. Chudnovsky 1991. *Proceedings ASME Meeting*: 101.

Kadota K. & A. Chudnovsky 1992. *Polymer Engineering Sci.* 32: 1097.

Kadota, K., S. Chum & A. Chudnovsky 1993. *J. Appl. Polymer Sci.* 49: 863.

Kunin, B. 1991. *Ph.D. Thesis*. Illinois: University of Illinois at Chicago.

Lu, X. & N. Brown 1986. *J. Polymer Sci.* 21: 2217.

Lu, X. & N. Brown 1986. *J. Polymer Sci.* 21: 4081.

Lu, X. & N. Brown 1990. *J. Material Sci.* 25: 29.

Lu, X. & N. Brown 1990. *J. Material Sci.* 26:612.

Lu, X., R. Qian & N. Brown 1991. *J. Material Sci.* 26: 917.

Popelar, C.F., C.H. Popelar & V.H. Kenner 1990. *Polymer Eng. and Sci.* 30: 577.

Popelar, C.H., V.H. Kenner & J.P. Wooster 1991. *Polymer Eng. and Sci.* 31: 1693.

Schapery, R.A. 1975. *Inter. J. Fracture*. 11: 141, 369, 549.

Sehanobish, K., A. Moet & P.P. Petro 1985. *J. Mater. Sci. Letters*. 4: 890.

Stivala, S.S., S.H. Patel, A. Chudnovsky, A. Kim & Z.W. Zhou 1994. *Proceedings of the Soc. of Plastic Engineers Conference*. 3: 3290.

Stojimirovic, A. & A. Chudnovsky 1992. *Inter. J. Fracture*. 57: 281.

Stojimirovic, A., K. Kadota & A. Chudnovsky 1992. *J. Appl. Polymer Sci.* 46: 1051.

Williams, J.G. 1987. *Fracture Mechanics of Polymers*. New York: Halsted Press.

Weibull, W. 1951. *J. Appl. Mech.* 18: 293.

Zhurkov, S.N. 1965. *Inter. J. Fract.* 1: 311.

Ageing of Materials and Methods for the Assessment of Lifetimes of Engineering Plant, Penny (ed.)
© *1997 Balkema, Rotterdam, ISBN 90 5410 874 6*

Fatigue investigation of polycarbonate used for aircraft canopies

H. Abramowitz, T. Hentea, Y. Kin & Y. Xu
Department of Engineering, Purdue University, Calumet, Hammond, Ind., USA

ABSTRACT: Experimental fatigue analysis of 1/2-inch polycarbonate specimens was developed. Specimens were fabricated from structural polycarbonate sheets which are used in aircraft canopies. Experiments were conducted at ambient, cool and elevated temperatures. The stress-life approach was used to determine some fatigue characteristics of the polycarbonate investigated. In addition, the fracture mechanics approach was used to develop an inspection procedure for failure prevention of parts during their service.

1 INTRODUCTION

Air transparencies are high life cycle cost items for the Air Force that necessitate frequent replacement. One of the failure modes encountered most frequently in the field is polycarbonate fracture. There are, for example, complaints of transparency failures in flight. The nature of failures is not quite clear, but some evidence implies that transparency life is limited by fatigue of polycarbonate ply (Kelley, 1986).

The transparency is often manufactured from a laminated composite material. Components of the composite are an acrylic fact ply a polycarbonate ply, interlayers and coatings. Design of the canopy allows to unload an acrylic ply and that is why the structural polycarbonate ply of the composite was the primary concern during fatigue investigation in this project.

2 MATERIAL AND SPECIMENS

The testing coupons were cut from 0.5-inch polycarbonate sheets. The sheets had been extruded, pressed and polished in accordance with the military specification. The specimens were cut by a fine band saw using the lowest possible speed and a cooling liquid. Three configurations of specimens (Figure 1) were investigated: solid, with holes as stress concentrators and with "V" notches.

The 3/4-inch core-series 17-0310 scotch tape was used to protect the gripping area of the specimens from damage. The scotch tape was bonded in three layers on each end of the specimen. It took roughly 10 minutes after the beginning of the test to

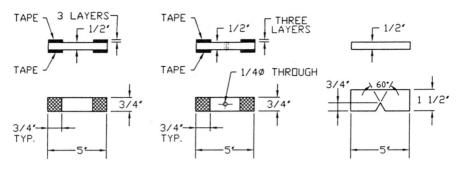

Figure 1. Three types of specimens used in study

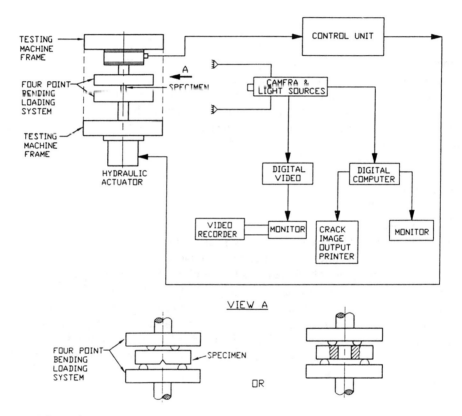

Figure 2. Experiment set-up

adjust the assigned loading regime due to formation of a "bed" by the hard jig rollers in the soft tape layers. The bed also prevented sliding of the specimen between the rollers.

3. EQUIPMENT

The flexure fatigue tests were conduced on an MTS machine using a four point MTS flexure system to provide pure bending. Experimental set-up is shown in Figure 2.

The appropriate support and load spans were selected to provide minimum possible deflection of the specimens. The small deflections provided more stable position of the specimens between jig rollers and permitted the use of a greater testing frequency. The flexure fatigue tests under different temperatures were conducted in the MTS environmental chamber.

The fatigue crack propagation images were caught by an image grabber linked to the digital computer and digital video. An independent duplicate monitor connected to the digital video was permitted constant observation of the crack development during some necessary command and control operations on the monitor connected to the computer.

4 FAILURE CRITERIA AND TEST PROCEDURES

Different options (certain percentage load drop, complete separation, certain percentage of crack propagation, crazing and visibility lost, crack initiation) were analyzed prior to the assignment of a failure criteria. The final decision was made to perform the test under constant load and stress controller regime until complete separation. The other failure criteria were also checked and showed more inconsistency and greater scatter (Kin 1994). The pulsating bending S-N tests were conducted on specimens exposed to temperature changes in the range from -50°C to 60°C.

The ratio of minimum load over maximum load was 0.2 for all tests. The testing time per day was not more than 10 hours, hence the possible influence of stops was not considered in this project. The tests continued until complete separation or not longer than 10^6 cycles. Regression analysis was used to treat the test results.

During fatigue crack propagation investigations, the crack images were captured by an image grabber after each 5000 cycles and stored. finally N-N, fatigue crack length-number of cycles and da/dN-ΔK diagrams were developed.

5 STRESS-LIFE TEST RESULTS (ROOM TEMPERATURE)

The test results of 0.5-inch coupons are plotted in Figures 3 and 4. The cracks were always started at the bottom tensile zone of the specimens. In all solid specimens with the exception of two cases the cracks propagated from the edge toward the center of the coupons tested. In all specimens with stress concentrators the cracks propagated from the hole edges toward the specimens sides. usually craze (minute cracks) spot cracks preceded the crack formation and propagated ahead of the crack tip. Thus it can be concluded that the damage mechanism is very similar to that described for crack propagation in polystyrene under fatigue loading (Chudnovsky 1987). When the visible separate minute cracks were detected during the high load level testing, the massive craze zone developed very fast. It can be noted that the lives of specimens are significant after the massive craze spot formation until complete breakage. For low load levels no massive craze zones were observed. For many of these cases also, after the initiated crack was easily visible, a substantial number of cycles were completed until breakage of the specimen. Comparison of the S-N diagrams in Figure

3 and 4 show that the stress concentration influence on fatigue life of the tested polycarbonate specimens is significant.

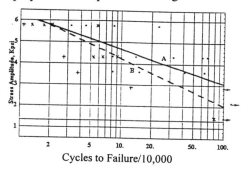

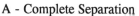

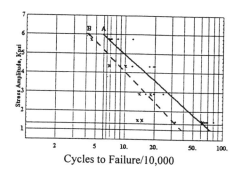

Figure 3. N-N Diagram for the Polycarbonate Specimens

Figure 4. S-N Diagram for the Polycarbonate Specimens with Holes

A - Complete Separation
B - Crack Initiation
o - Failure
x - Crazes
+ - Specimens with not Polished Edges

A - Complete Separation
B - Crack Initiation
o - Failure
x - Crack Initiation

6 TEMPERATURE EFFECTS

Flexure fatigue resistance of the 0.5-inch polycarbonate coupons exposed to temperature changes in the range from -50°C to 60°C was analyzed. The test procedure, specimen design, equipment, loading type and regimes were the same as described for the test under ambient conditions.

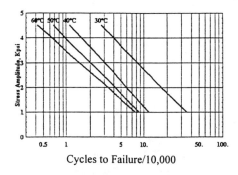

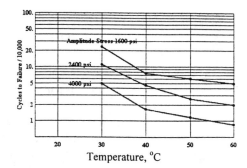

Figure 5. S-N Curves for the 0.5-in
Polycarbonate Specimens
Under Different Temperatures

Figure 6. Temperature Effects on Fatigue
Fatigue Life of the 0.5-inch
Polycarbonate Specimens

The influence of increased temperature (Figures 5 and 6) is very strong and the fatigue lives of the specimens tested are from 2 to 3 times shorter compared to the test results under room temperature. It is important to note that complete fractures of

312

Table 1. Results of 0.5-inch polycarbonate coupon fatigue tests under cold temperature.

Specimen number	Amplitude stress, psi	Temp., C°	Number of test cycles	Notes
1	6750	-50	119000	Failure
2	6750	-50	132000	Failure
3	4000	-50	100000	No Failure
4	4000	-50	100000	No Failure
5	4000	-50	100000	No Failure
6	2400	-50	100000	No Failure
7	4000	-5	100000	No Failure
8	2400	-5	100000	No Failure
9	2400	-5	100000	No Failure

Table 2. Comparison of the Fatigue Lives of the 0.5-inch Polycarbonate Specimens Tested under Room, ** Cold and Elevated Temperatures.

Amplitude stress, psi	Average number of cycles until failure					
	+30°C	+40°C	+50°C	+60°C	-50°C	-5°C
1600	220,000	80,000	60,000	40,000		
2400	110,000	42,000	26,000	20,000	100,000*	100,000*
4000	45,000	15,000	11,000	8,500	100,000*	100,000
6750	-	-	-	-	125,000	

*   No Failure
** Laboratory temperature was 30°C

the coupons tested under increased temperatures followed almost immediately after crack initiations. Therefore, there was no "warning" before the break. Conversely, a very significant increase of fatigue resistance under cold temperatures was observed. The results of 0.5-inch polycarbonate coupon fatigue tests under cold temperatures are given in tables 1 and 2. The increase was so significant that we can assume that the investigations of the selected material under cold temperatures are not critical for the application considered in this project.

7 CRACK GROWTH RESULTS

From the results of stress-life investigation described it can be noted that for many cases the fatigue life after crack initiation is still very significant. Therefore it is important to determine crack growth parameters. On the basis of crack growth characteristics the part expected life still left can be estimated and the appropriate inspection intervals assigned. The plots of fatigue crack length versus number of cycles for the specimens tested under different loads are given in Figures 7, 8 and 9 and corresponding crack growth rate curves are given in Figures 10, 11 and 12.

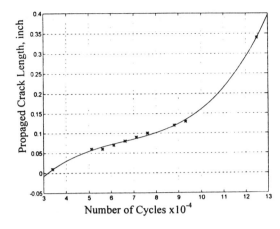

Figure 7. Crack length (from the tip of the notch) versus number of cycles. Max. load = 400 lb., Min. load = 100 lb.

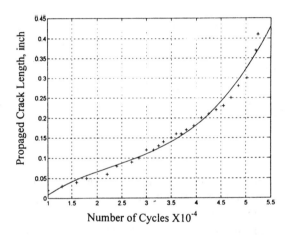

Figure 8. Crack length (from the tip of the notch) versus number of cycles. Max. load = 450 lb., Min load = 100 lb.

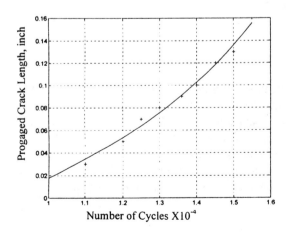

Figure 9. Crack length (from the tip of the notch) versus number of cycles. Max. load = 500 lb., Min load = 100 lb.

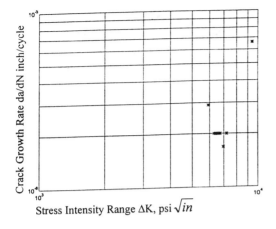

Figure 10. Crack growth versus stress intensity range. Max. load = 400 lb., Min load = 100 lb.

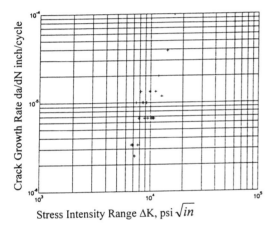

Figure 11. Crack growth rate versus stress intensity range. Max. load - 450 lb., Min. load = 100 lb.

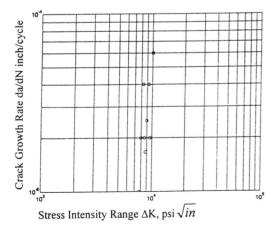

Figure 12. Crack growth rate versus stress intensity range. Max. load - 500 lb., Min load = 100 lb.

315

The crack growth rate da/dN was calculated by a polynomial approach which fits a cubic equation to a certain interval of the crack length versus number of cycles curve. Stress intensity factors were determined (Julie et al 1990), (Brown & Stawley 1996) for a cracked beam in pure bending

$$K = f(g)\sigma\sqrt{\pi a}$$

where a = notch height (see Figure 1)
σ = bending stress, determined from given load, location of load application and specimen geometry

$f(g)$ = correction factor

$$= 1.122 - 1.40\,(a/b) + 7.33\,(a/b)^2 - 13.08\,(a/b)^3 + 14.0\,(a/b)^4 \quad \text{(Julie et al 1966)}$$

Using the Paris formulation, for example, the life to failure can be determined during a maintenance inspection.

$$N_f = \int_{a_i}^{a_f} \frac{da}{C(\Delta K)^m} :$$

N_f = life to failure
b = specimen height
a_i = initial crack length
a_f = final (critical) crack length
ΔK = stress intensity range
Parameters C and m for the specimen tested were determined from test results. Values for m were from 2.17 to 3.057
Values for C were from $e^{-39.927}$ to $e^{-36.987}$

CONCLUSIONS

The 0.5-inch polycarbonate coupons have substantial fatigue life after crack initiation until complete failure.

The stress concentration effect on the fatigue life of the polycarbonate tested is very strong.

The fatigue life of the polycarbonate is significantly decreased as the temperature is increased (from 20°C to 60°C).

Relationships between fatigue crack length and number of cycles, crack growth and rate and stress intensity factors were obtained for the material investigated. Based on these results recommendations to maintenance policy can be developed in the form of software.

ACKNOWLEDGEMENT

Partial sponsorship by the Flight Dynamics Directorate, Wright Laboratory, Wright Patterson Air Force Base, Dayton, Ohio, is gratefully acknowledged.

REFERENCES

Bannantine, J.A., Comer, J.J., Handrock, J.L. 1990. *Fundamentals of Metal Fatigue Analysis*. Prentice Hall, Englewood Clifts, New Jersey.

Brown, W.F. and Srawley, J.E. 1996. *Plain Strain Crack Toughness Testing*. ASTM STP410, American Society for Testing and Materials, Philadelphia, Pennsylvania.

Chudnovsky, A. Chaovi, K., Moet, A. 1987. "Curvilinear Crack Layer Propagation", *Journal of Material Science Letters 6*, pp 1033-1038.

Kelley, M. FIER, 1986. *Examination Report Wright-Patterson AFB,* Dayton, Ohio

Kin, Y.B. 1994. "Fatigue Failure Analysis of Polycarbonate Transparencies in Different Environmental conditions". *Final Report Flight Dynamics Directorate*, Wright Patterson AFB, Dayton Ohio.

Author index

Abramowitz, H. 309
Aladinsky, V.A. 97
Aladinsky, V.V. 63
Andersson, P. 129
Arzhaev, A.I. 63, 97

Badalyan, V.G. 97
Baron, D. 297
Beattie, A.W. 193
Bougaenko, S.E. 63

Chrzanowski, M. 117
Chudnovsky, A. 297
Clarotti, C.A. 85
Collins, J.A. 147

Danilov, V.L. 107, 113
Davies, M.A. 71
Delamarian, C. 41
Denisov, I.N. 63
Dobrov, M.V. 107

Fautrelle, Y. 107
Furtado, H.C. 147

Hentea, T. 309
Hernas, A. 181
Hsu, K.Y. 273

Irvine, N.M. 39

Jakowluk, A. 229
Jeon, J.-Y. 219

Kautz, H.R. 41, 167, 261
Kin, Y. 309
Kiselyov, V.A. 97
König, H. 157, 241
Konvičková, S. 51

Lannoy, A. 85
Latus, P. 117
Le May, I. 147
Lee, J.-Y. 219
Levit, V.I. 207, 285
Lilley, J.R. 71

Makhanev, V.O. 63, 97
Mayer, K.H. 157, 241
Meslin, Th. 95
Michalec, J. 51
Mirecki, L. 181
Molyneaux, M.J. 23

Procaccia, H. 85

Řezníček, J. 51
Růžek, L. 17

Růžička, M. 51

Samuelson, L.Å. 129
Schnack, E. 3
Segle, P. 129
Shulkin, Y. 297
Španiel, M. 51
Sochor, M. 51
Speck, J.B. 23
Storesund, J. 129
Strelkov, B.P. 97

Tipping, Ph. 285
Tomkins, B. 39

Valenta, F. 51
Vanukov, V.N. 97
Vejvoda, S. 17
Vincour, D. 17
Vopilkin, A.Kh. 97

Weber, D. 157
Weber, H. 251
Weiss, M. 157

Xu, Y. 309

Zarubin, S.V. 107, 113
Zschau, M. 251